MINING EXPLAINED

A Layperson's Guide

John Cumming
Editor

Doug Donnelly
Publisher

THE
NORTHERN MINER

2012

Library and Archives Canada Cataloguing in Publication

Mining explained : a layperson's guide
    John Cumming, editor — 11th ed.

Includes index.
ISBN 978-1-55257-147-7

1. Mineral industries–Canada.
2. Mining engineering–Canada.
3. Prospecting–Canada.
I. Cumming, John, 1968-

TN147.M5 2007   622'.0971   C2006-906996-4

A publication of THE NORTHERN MINER
Doug Donnelly, Publisher
John Cumming, Editor
80 Valleybrook Drive
Toronto, ON Canada
M3B 2S9
www.northernminer.com

MINING EXPLAINED

Table of Contents

1. Introduction

THE MINING WORLD

There are many thousands of producing mines in the world and more are being developed every year. From out of the ground, they provide the raw materials for the manufacturing, construction and chemical industries and most of the energy minerals on which a modern, technological society is so dependent.

These mines produce many of the things we often take for granted. They are the source of all the metals you see in the buildings, cars, airplanes, electronic gadgets and household products around you, many of the raw materials for our building and chemical industries and even the coal and uranium which produces much of the electricity we consume every day. Even the strengthening agent in the paper on which this book is printed is derived from the product of a mining operation somewhere in the world.

Although mining in Europe is in decline, the historical mining areas of Saxony and Silesia in eastern Europe and the tin fields of Cornwall in England are cradles of the technologies of mining, mineral processing and metallurgical processes.

The Mediterranean countries of Greece, Italy, France, Spain, Portugal and Turkey have had their histories much coloured by the earliest of mineral production and trade. The Greeks financed their wars with the Persians largely with silver produced at Laurium, and the Romans produced copper at Rio Tinto, an area still in operation. The Phoenicians carried on much of this trade and travelled as far from their homes as Cornwall for tin.

The search for metals, precious stones and other minerals is at the root of many of the great events in history, both ancient and modern. This process has determined much of the world's economic geography and demography. Many of the world's developing countries export their mineral wealth to the developed world in exchange for foreign currency, making minerals an important part of international trade.

Mining is key to the great wealth of Canada, and was one of the principal reasons for the opening of the North to modern development and transportation. The Sudbury nickel district, the gold camps of the Abitibi, the iron ranges of Labrador, the potash mines of Saskatchewan and the mines of the British Columbia interior are only a few of the great mining areas of Canada. Today, there are about 130 mines in Canada, both open-pit and underground. They produce about 3% of Canada's gross domestic product and provide over 19% of its goods exports.

The United States has abundant mineral resources, notably in Nevada, which is now one of the world's busiest gold mining areas. After more than a century, Idaho continues to be a world-class silver district. America also has the great copper camps of Montana, Utah, Arizona and New Mexico and the lead and zinc of Missouri. Nebraska and Wyoming produce uranium, while Minnesota, Wisconsin and Michigan produced iron ore that fueled the Industrial Revolution.

The history of much of Latin America and Australia, too, has been shaped by minerals, while many Asian and African nations will be key mineral producers for decades to come.

Mining is as old as civilization itself – as old as the hills, one might say – and with each new use for the Earth's minerals discovered in this hi-tech age, it becomes clearer that mining will always be with us.

Our Goal

This book has been written for anyone with an interest in mining. It is the eleventh edition of *Mining Explained* to be produced by *The Northern Miner* since 1939. This edition has been revised and edited by John Cumming.

The scientific, engineering and fiscal processes employed in the transformation of a mineral prospect to a commercial-scale producer are all here, but simplified and spelled out in plain English.

This is no textbook; it is simply a guide for those readers of *The Northern Miner* and others who are interested in knowing how a mineral is found, mined, processed and turned into cash.

The glossary is meant to serve as a companion piece to each week's issue of *The Miner*, shedding light on the myriad geological and financial terms found in the pages of the newspaper.

Whatever your area of interest, *Mining Explained* has been designed to provide you with a greater understanding of the world of mining. We hope you find that it lives up to its title.

2. Basic Geology

ROCKS AND MINERALS

Investors and the thousands of people employed by the mining industry don't need to know as much about the Earth as a geologist does. But they should know at least the basics of geology and have some knowledge of the more common rocks and minerals.

To the investor who "puts his money in the ground", a little knowledge of geology can go a long way. For instance, an investor might some day benefit from knowing that low-temperature mineral deposits, such as most silver orebodies, change quickly with depth or that faulting can play havoc with an orebody.

Geology, the science of the earth, has many branches. We'll deal here only with those that have a direct bearing on finding mines, beginning with a discussion on how mineral deposits form in the Earth's crust.

The five most abundant elements in the crust are oxygen, silicon, aluminum, iron and calcium. Elements bond together in chemical compounds

to form solid crystalline substances known as minerals. (Non-crystalline mined materials such as coal are also often called "minerals" in an economic, if not a scientific, sense.)

There are many thousands of different minerals, each with a definite chemical composition and crystalline structure. For instance, the elements iron and sulphur combine in a definite ratio to form the yellowish mineral pyrite. The elements potassium, aluminum, silicon and oxygen combine in a definite ratio to produce the common rock-forming mineral orthoclase.

The common rock-forming minerals are almost all silicates – that is, minerals containing silicon and oxygen. Sulphides are minerals with elements chemically bonded to sulphur, whereas oxides have elements bonded to oxygen. The fourth important mineral group is the carbonates, in which carbon and oxygen are bonded with other elements.

What is the difference between a rock and a mineral? A rock is simply a solid mass of mineral grains. A rock is classified according to the kinds and proportions of minerals it contains.

Where minerals are concentrated in sufficient quantity, the zones or bodies in which they are found are called **mineral deposits**. Mineralization becomes **ore** when the minerals are present in sufficient quantity, or tonnage, and adequate quality, or grade, to be recovered *profitably*.

These deposits are often mined to produce metals or other commodities such as coal and uranium oxide. Most people can easily identify commonly used metals such as gold, silver, copper, zinc, aluminum and tin. These metals have a distinctive lustre, conduct heat and electricity, and are malleable under heat and pressure.

An **alloy** is a metallic material consisting of two or more elements, homogenous in outward appearance, and combined in such a way that they cannot be readily separated by physical means. Alloys are used more often than pure metals because they frequently have properties different than those of the metals from which they are made. For example, combining copper and tin creates an alloy that is harder and tougher than either of the two metals in their pure form. The alloy is bronze, and its discovery marked the period of civilization known as the Bronze Age. Today, alloys are used in the advanced technologies of aerospace, electronics and energy.

Metals are often categorized into groups to reflect some common uses or similar properties. Precious or noble metals include gold, silver and platinum, while base metals are those of lower value, mainly copper, nickel, lead and zinc.

Ferrous metals are those with a strong chemical affinity to iron, and are often used in steelmaking. Chromium, cobalt, manganese and molybdenum are commonly included in this group, sometimes called ferroalloy metals, because their major use is to improve the properties of steel.

Non-ferrous metals include aluminum, copper, lead, magnesium, nickel, tin and zinc, since their largest uses are not in steel-making (there is some overlap here with the base metals).

Light metals such as beryllium, magnesium, aluminum and titanium are valued for their combination of lightness and strength. Refractory metals such as niobium, tungsten and ruthenium have high melting points and can withstand high temperatures. Reactive metals such as lithium, strontium and cesium are less stable because they easily react with oxygen. Nuclear or radioactive metals include radium, thorium, uranium and plutonium. They are often used to generate power.

Semi-metals or **metalloids** such as silicon, arsenic and selenium possess a mix of metallic and non-metallic attributes. Another group is rare-earth elements (REEs) and includes scandium, yttrium, zirconium and fifteen elements collectively termed the lanthanides. These REEs are so categorized not because they are rare in the earth's crust, but because their extraction is difficult.

Metals and minerals are a natural part of our environment. Finding economic concentrations of them in the earth's crust is a task that has preoccupied humankind since the dawn of civilization.

ELEMENTARY GEOLOGY

The minefinder needs to understand the physical and chemical processes that make the Earth produce its mineral wealth. For example, understanding plate tectonics – formerly called "continental drift" – allows the geologist to locate areas of plate collision where seismic and geological activity can form mineral deposits.

Earth is very old indeed; the oldest known rocks date back nearly four billion years. And geological processes move slowly; a mountain range several kilometres high is lifted at a rate of a few centimetres per year. Erosional forces wear away such a mountain range just as slowly.

In many parts of the world, there have been not one but many such cycles of uplift and erosion. Periods of mountain-building are referred to as orogenies. The east coast of North America, for example, has undergone at least three distinct periods of mountain-building, each corresponding with the opening and closing of what is today the Atlantic Ocean.

The most recent opening started about 200 million years ago and is still going on. A more recent orogeny created the mountain ranges on North America's west coast.

COOLING PROCESS IMPORTANT

The Earth has a solid core of iron and nickel surrounded by a mantle of molten rock. When this material forces itself into the many cracks and other points of weakness in the crust, it is called magma. These tongues of molten rock, which move out in many directions, heat the surrounding rock, altering it and, in some cases, causing it

to re-melt. The whole mass then cools and that is when minerals, some of which are valuable, begin to crystallize.

Mineral constituents of magma crystallize at different temperatures. As a result, they tend to concentrate during the cooling process. It is these concentrations that often form mineral deposits. As each different mineral crystallizes out of the magma, the composition of the magma changes. Heavier minerals sink to the bottom of the semi-liquid mass and are concentrated in a process called magmatic segregation. Some magmas have no valuable minerals, while others contain economically exploitable mineral deposits.

Other deposits may be formed by minerals dissolved in the circulating liquids in and around magmas; these are redeposited at some place where chemical and physical conditions permit. These processes are known as hydrothermal processes.

When hot, sulphide-laden fluids spew up through fractures in the sea floor, these fluids form layered volcanogenic deposits. These are an important source of zinc, copper, lead and gold.

The processes of weathering can also produce economic concentrations of minerals, for example, placer gold found in river beds.

Because no two orebodies are alike, comparison among mines is risky. Just because ore on a particular property continues to depth does not mean that ore on an adjoining property will follow the same trend.

IGNEOUS ROCKS

Igneous rocks are those that have been formed by the cooling and crystallization of a magma. All magmas originate at great depths in the Earth's crust, where both temperature and pressure are very high. When opportunity permits, magma will expand and flow toward surface. The magma may cool and solidify below surface, allowing individual mineral grains to grow together, forming an interlocking pattern.

Rocks that cool at great depth do so slowly, giving crystals time to grow large. This coarse-grained texture is typical of these intrusive igneous rocks. Molten rock that comes closer to surface will cool faster, so the crystals will have less time to grow. The resulting rock is finer grained.

Some magma may reach the surface, cooling quickly to form a lava flow. The resulting rock is very fine-grained and called a volcanic, or extrusive, rock. Magma containing trapped gases can also erupt with an explosive force from a vent or volcano.

The material ejected from volcanoes varies in size from fine dust to rock fragments weighing hundreds of tonnes. This includes both lava and solid or semi-solid rock from around the vent or the volcano's crater. When this material settles, the larger angular pieces form rocks known as agglomerate or volcanic breccia, while the finer materials form tuff.

The composition of an igneous rock depends on the composition of the magma that produced it. If the magma is high in iron and magnesium, the resulting rock will have abundant iron and magnesium silicate minerals and will be dark in colour; these rocks are called mafic rocks. A magma high in silica will form a light-colored rock with abundant quartz and feldspar; such rocks are called felsic rocks. There are also rocks with compositions intermediate between these two extremes.

Intrusive rocks made up wholly of dark minerals – rocks like dunite and peridotite – are termed ultramafic. As the proportion of dark minerals decreases, the rocks are called gabbro, diorite, granodiorite, and granite, with granite being the most felsic. Syenite is similar to granite, but has less silica.

Volcanic rocks are classified in the same way as intrusive rocks, so that each intrusive rock has a volcanic counterpart with a similar composition. Komatiite is the ultramafic volcanic rock, and like peridotite, contains only dark minerals. Basalt corresponds to gabbro, andesite to diorite, dacite to granodiorite and rhyolite to granite. Trachyte is the volcanic equivalent of syenite.

An igneous rock that contains distinct crystals embedded in a much finer-grained groundmass is called a porphyry. The name applies only to this characteristic texture, not to the chemical composition of the rock. The groundmass of a porphyry can have either a volcanic or intrusive texture. Porphyries are usually given a compound name signifying their composition, as in dacite porphyry. Similarly, volcanic fragmental rocks are usually given compound names like rhyolite tuff.

SEDIMENTARY ROCKS

Sedimentary rocks fall into two categories, clastic and chemical, based on their origins. Clastic sedimentary

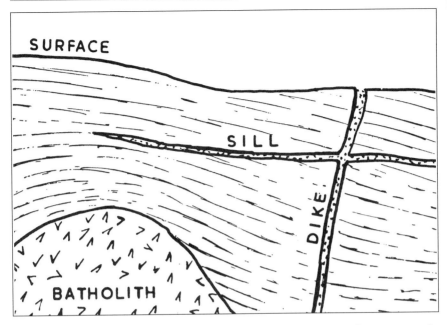

SURFACE

SILL

DIKE

BATHOLITH

A large body of intrusive rock extending to great depth is a batholith. Sometimes these intrusive rocks take the form of dykes and sills.

rocks are consolidated fragments of eroded rock, brought together by the action of ice, water and wind. In a conglomerate, large rounded rock fragments are held together by a groundmass of finer fragments. Sandstone, as the name implies, is made up of sand grains, although it can also contain finer material. When most of the material is fine-grained, the rock is called siltstone or shale.

Chemical sedimentary rocks form when dissolved materials precipitate, the chemist's term for falling out of a solution as a solid material. The most common is the carbonate rock limestone, composed of fine-grained calcium carbonates often derived from the shells of sea creatures. Dolostone is a carbonate rock made up of dolomite rather than calcite. Chert is a chemical sediment made up mostly of quartz. Iron formations are composed of iron oxides, sulphides, carbonates or silicates.

Chemical sediments can also form where ocean water becomes isolated from the ocean. These rocks, called evaporites, can contain gypsum (a calcium sulphate), halite (rock salt) and sylvite (a common potash salt).

Bedding is a common characteristic of most sedimentary rocks. In these rock types, it is usually possible to see the individual layers of material as they were laid down over time. Beds start as flat layers, but can be folded or overturned by later tectonic forces. Coarser particles typically form a distinct band, whereas finer particles form progressively higher bands or beds. However, the individual particles in sedimentary rocks do not show the same interlocking features of igneous rocks. The individual grains in sedimentary rocks are actually cemented together.

METAMORPHIC ROCKS

Metamorphic rocks are really igneous or sedimentary rocks that have undergone extreme physical and chemical changes. In most cases, these changes are so severe that the rocks' original identities are obscured.

If subjected to high temperatures or pressures, the original constituents of a rock can be transformed into new minerals that are chemically stable at these higher temperatures and pressures. These new mineral particles readjust themselves into parallel and flattened patterns to conform to the pattern of pressure affecting the rock. It is this laminated or banded texture that characterizes most metamorphic rocks. While at first glance this banding might suggest a sedimentary origin, closer examination will usually reveal a distinct interlocking texture, unlike the cemented texture of a sedimentary rock.

A common and useful concept is metamorphic grade. This expresses the degree to which metamorphic heat and pressure have changed a rock. Low-grade metamorphism typically changes the mineral composi-

tion only slightly. At progressively higher grades, more and more of the original minerals react chemically with each other, recrystallizing to form new minerals, and releasing water, carbon dioxide and other volatile constituents. At the highest metamorphic grades, the volatiles are driven off completely and the rock may begin to melt.

Frequently, metamorphic rocks are named for their parent rock types. For example, a metamorphosed volcanic rock may be called a metavolcanic or a metamorphosed sedimentary rock a metasediment.

Gneiss (pronounced "nice") is the name of a common group of metamorphic rocks. When used alone, the term generally signifies a rock displaying the distinctive laminated feature of high-grade metamorphic rocks. Such rocks are typically composed of quartz, feldspar and mica. Squeezing a granite will cause the light and dark crystals to be dragged into more or less parallel bands. It is then called a granitic gneiss.

The common medium-grade meta-

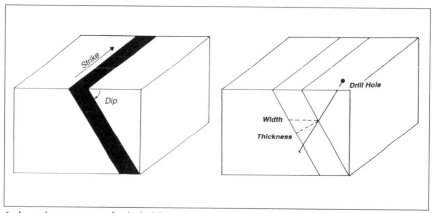

Strike is the orientation of a bed, dyke or vein at surface; dip is the angle the body makes with the horizontal. Unless a drill hole intersects a bed at right angles, the length of the core intersection will be greater than the true thickness of the bed.

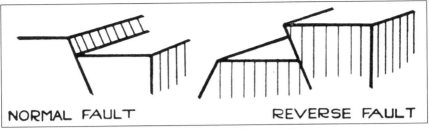

NORMAL FAULT   REVERSE FAULT

On a normal fault, the motion is down-dip. On a reverse fault, it is up-dip.

morphic rock type is schist. The rock has a laminated structure in which the recrystallized minerals have been oriented in parallel bands. This rock tends to break along its layers, unlike gneiss, which is more completely recrystallized. A schist is usually named for the most common minerals it contains, like talc schist, mica schist and chlorite schist.

Shale, a common sedimentary rock, when subjected to high pressures, changes into a rock called slate. This rock is often used as a roofing material as well as a base for pool tables.

Quartzite is an extremely hard rock formed by the metamorphism of pure quartz sandstone.

Marble is another metamorphic rock, formed from limestone which has been recrystallized. It is used extensively as a building stone.

Rock Formations and Structures

Rock bodies come in all shapes, but the shapes are all variations on a few basic forms. Sedimentary and volcanic rocks all start out as flat layers or beds, which may later be deformed by the continuing motion of the Earth's surface. If the beds are tilted, the geologist describes the orientation of the bed in terms of its strike and dip.

The strike and dip of an orebody can be imagined as a sheet of ply-wood sitting on an angle in a tank of water. The line where the sheet and the surface of the water intersect is the line of strike – a horizontal line with a bearing that can be measured with a compass. The angle the wood makes with the surface of the water is its dip. Both strike and dip are usually measured in degrees.

Folds are caused by sideways pressure on flat-lying bedded rocks which buckle and form a wavy pattern. Bending this book slightly, and viewing the top edges of the pages gives an idea of the structure of a fold. Folds with "crest" shapes are called anticlines, and those with "trough" shapes are called synclines.

Intrusive rocks generally form large bodies called batholiths or smaller, pipe-like bodies called stocks. A dyke is a sheet-like intrusive body that cuts through the surrounding country rock. A sill is also sheet-like, but it forms along a space between bedded rocks. Like bedded rocks, dykes and sills have a strike and dip.

Fractures are very common in rock. If the fracture is large enough, and the rocks have been torn apart, the fracture is called a fault. If there is intense movement along a fault, it is called a shear zone.

Fractures, faults and shear zones, like beds and dykes, are planar features that can be described by their strike and dip. It is also useful to label the upper surface of an inclined fault the hangingwall, and the lower surface the footwall.

It is important for the geologist to know the sense of movement along the fault – upward, downward or sideways – because this allows rock units broken apart by the fault to be traced. For example, if an orebody is broken apart by later faulting, knowing whether the hangingwall has moved up or down against the footwall allows the geologist to predict where the rest of the orebody might be.

Fracturing can also help to form orebodies. Fractures and openings in the rock allow fluids to pass through, and dissolved material in the fluids may be left behind to form a vein.

Some structures have a more linear shape – for example, a pipelike intrusive body, the axis of a fold, or an ore shoot in a vein. A linear structure's orientation is better described by its trend, which is the direction it heads, and its plunge, which is the angle it makes with the earth's surface. It is common to speak of a structure as "plunging gently to the southwest" or "steeply to the north", which means only that it descends in that general direction; for more precise applications, the trend and plunge are expressed in degrees.

OVERBURDEN

Often overlying the bedrock of the earth's crust is a blanket of soil, gravel, clay and other unconsolidated material collectively called overburden. It is formed by the weathering action of wind, rain, runoff water, groundwater, and even by the growth of plants and the burrowing of animals.

Soil is the weathered material that develops as bedrock – or sometimes other forms of overburden – breaks down in place. A rock body undergoes significant chemical changes when it turns into soil, with hard silicate minerals turning into soft clays and oxides, and with "dry" minerals taking on water from their surroundings.

The most profound weathering results in saprolites and laterites. These are unconsolidated materials that form from solid rock under tropical weathering conditions. Saprolite is a strongly decomposed rock that has mainly been converted to clay minerals; some of the most resistant silicates, like quartz, may remain. In a laterite, only very resistant minerals remain, with the rock undergoing an almost total loss of silica; true laterite consists mainly of secondary iron and aluminum oxides and hydroxides.

In mountainous country, mechanical weathering (the simple action of gravity) assumes greater importance. Mountain valleys are typically edged with scree, a mass of broken and abraded rock and sand particles formed as slopes and cliffs collapse. The valley floors will usually be filled with alluvium.

Terrains that have been glaciated will have a wide variety of overburden materials. Till is the mixture of boulders, gravel, sand and silt smeared across the landscape by a continental glacier. Other material, collectively called glacial drift, is emplaced by sedimentary processes as the glacier melts.

3. Ore Deposits

Ore Deposits: The Big Picture

While there is certainly an element of luck in the discovery of any orebody, locating one is usually more than just the result of fortuitous circumstances. The formation of an orebody calls for special conditions that need to be understood by the mine finder.

Looking at the history of mining around the world, it is the areas of Precambrian shield rock, along with the much younger mountain belts of western North and South America, Africa, Asia and Australia that have yielded most of the world's mines.

The Canadian Shield, occupying almost the entire central and northeastern portions of Canada, is one of the most prolific mining areas of the world – from the iron mines of Labrador to the gold and base metal mines of Val d'Or, Que., and Timmins, Ont., all the way northwest to the diamond mines near Yellowknife, N.W.T.

The Canadian Shield is composed of rocks formed during the Precambrian era (from 4.6 billion to 570 million years ago) and is one of the cooling Earth's original land masses. Many of these rocks have since been highly metamorphosed. This region, as well as the Precambrian shields on every continent, have proven to be rich with metal orebodies, mostly because the mountain-building and other tectonic activity that leads to the formation of orebodies was widespread and intense in these early days of the planet's life.

Hundreds of millions of years of erosion have since worn the old shield mountains down to their roots, bringing their ore nearer the surface.

The Cordillera, a band of relatively young and unstable rocks, covers an area that stretches from Alaska to British Columbia and the Pacific states down into Mexico and Central America and the Andes. It features complex rock structures favourable for metallic ore deposits, hosting some of the largest base metal and richest gold deposits ever found.

More than half a billion years after much of the tectonic activity in the Earth's shields ceased, the Cordilleran region became as active as the shields once were, leading to the formation of some of the world's biggest ore deposits in places like British Columbia, Nevada and Chile.

How Ore Forms

One useful way to classify mineral deposits is to distinguish between those that were formed at the same time as the host rocks from those that were formed afterward. Syngenetic mineral deposits are those which form from igneous bodies or by way of sedimentary processes. Epigenetic mineral deposits form in rocks that already exist: for example, solid rock may fracture and veins may be deposited later in the fractures, forming an epigenetic deposit.

Ores can be formed by the processes that produce rocks – there are mineral deposits that appear to have been created by the crystallization of a magma or from the erosion and redeposition of material that comprises sedimentary rocks. But mineral deposits also form by another process, called hydrothermal activity, which is the action of heated fluids in the earth. Many mineral deposits are chemical precipitates from hydrothermal solutions – that is, they have come out of solution as solids.

In a hydrothermal process, hot water, circulating through rocks by way of fractures and pore spaces, can leach minerals out of the rocks through which it passes and transport the minerals in solution. The minerals remain dissolved in the water until something makes them precipitate. A number of things can happen to do this. Sometimes the temperature falls or the confining pressure of the rock suddenly decreases. Other times the water encounters another rock type that reacts chemically with the dissolved metal, forming new minerals. Sometimes one fluid meets another with different chemical species in solution, and the dissolved species from each fluid react.

A mineral deposit is made up of ore minerals, which carry the metal, and gangue minerals, which are formed along with the ore minerals but contribute nothing to the value of the deposit. For example, gold veins often are made up of large amounts of quartz and carbonate gangue, with some pyrite and a little gold. Only the gold is there in a form and amount that is worth extracting.

Wherever the hot water goes, it reacts chemically with the rock, causing alteration. Alteration is the chemical destruction of some or all of the existing minerals in a rock and the creation of new ones. Mafic minerals like pyroxene can be converted to chlorite; feldspars are converted to micas and clays; carbonate and sulphide minerals and quartz are left behind in the rock. Hydrothermal alteration is a sign that fluids have passed through a rock, and is one of nature's clearest

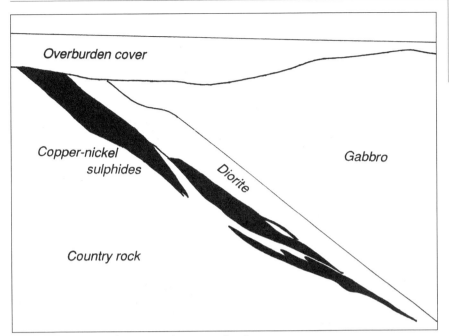

Copper-nickel sulphides often occur at the contacts of mafic intrusions.

messages that there may be a mineral deposit nearby.

We will now look at some common types of mineral deposits and their geological characteristics. The diagrams that accompany some of the text are cross-sections, showing the shape the deposits typically have below the surface of the earth.

MAGMATIC DEPOSITS

As a body of molten rock cools, minerals begin to crystallize, sinking to the base of the magma chamber or accreting to its sides. These are usually the ordinary rock-forming minerals, but sometimes they are useful minerals like base metal sulphides or the oxides of iron, titanium or chromium. Ultramafic or mafic igneous rocks can have concentrations of these minerals that are large enough to be minable.

Much of the world's nickel and a large proportion of its copper and cobalt come from sulphide deposits of this kind in mafic igneous rocks, like the nickel-copper deposits of Sudbury, Ont.

Platinum, palladium and several other precious metals are often recovered as a byproduct in nickel-sulphide deposits. The sulphides can be massive or disseminated through the silicate rock. Pyrrhotite and chalcopyrite, plus one or more nickel sulphides, are the common minerals. The mineralized bodies usually have tabular or a lens-like shape.

Titanium and chromium oxides can also form magmatic deposits, either massive or disseminated. The mineralized bodies can even be layered in appearance like sedimentary beds, although their origin is clearly not sedimentary.

Diamond Pipes

Diamonds are formed in the rocks peridotite and eclogite, 150 km or more below the Earth's surface, where extreme temperature and pressure make diamond, not graphite, the stable form of the element carbon. Diamonds are brought nearer the surface by volcanic pipes and dykes of kimberlite, a dark ultramafic rock consisting mainly of the mineral olivine. The kimberlite eruption carries xenoliths, or rock fragments, of diamond-bearing eclogite or peridotite up through the Precambrian basement rock and overlying rock formations.

Mushroom-shaped kimberlite pipes, the most common type, are typically found in clusters. Some may contain economic quantities of diamonds, some may contain only a few diamonds and others may not be diamond-bearing at all. There are a number of more numerous "indicator" minerals, such as certain types of garnets, spinels and ilmenite, that can reveal the presence of kimberlites.

Diamonds have also been found in dykes of lamprophyre, an ultramafic rock similar to kimberlite, and in volcanic breccias.

Commercial quantities of lode, or hard-rock, diamonds have been found in locations in sub-Saharan Africa, Siberia, Australia and Canada's Northwest Territories.

Stratabound Massive Sulphides

This is a morphological term for base metal sulphide deposits that occur as part of a sequence of volcanic or sedimentary rocks and conform to their host rock's bedding. That they occur as part of the sequence is strong evidence that they formed along with their host rocks, rather than being emplaced later. They may start with tabular or lens-like shapes, but later deformation can fold them into complex shapes or break them into pieces. The expression "massive" sulphide has nothing to do with size; rather the mineralized bodies are nearly homogeneous and made up almost entirely of sulphides.

Volcanogenic massive sulphides are stratabound deposits in volcanic rocks. Volcanic vent areas and the dykes, sills and stocks that feed them are sources of heat and, consequently, are centres of hydrothermal or exhalative activity. Circulating waters carrying dissolved metals travel through fractures in the volcanic rocks, sometimes depositing sulphide minerals in the fractures themselves. The heat forces the fluids upward to the top of the volcanic sequence, where they are "exhaled" or vented.

The sudden change in temperature as the fluids leave the hot rock makes it impossible for the dissolved species to stay in solution, so they precipitate at the surface of the volcanic pile. It is common for this process to be occurring at several places at once and several deposits may form at a single stratigraphic level or mineralized horizon.

If the volcanic pile is under water, the exhaled fluids mix with the seawater and sulphides are deposited around the hydrothermal vent, forming layers of massive sulphide mate-

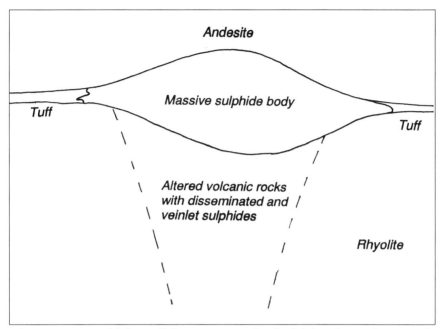

Volcanogenic massive sulphides usually have more copper than the shale-hosted types. There is usually disseminated mineralization in the footwall.

rial. They may be carried some distance by ocean-floor currents or may slump into ocean-floor depressions.

If the volcanic pile is above sea level, the fluids that were once under great pressure inside the fractures in the rock are suddenly released into the lower pressure of the atmosphere. The fluids respond by rapidly turning into steam. Dissolved minerals precipitate as a sinter around the vent area, much like scale around the mouth of a kettle.

Massive sulphide deposits can contain base metal sulphides like chalcopyrite, sphalerite and galena. The iron sulphides pyrrhotite and pyrite almost always occur along with them. The main ore minerals provide copper, zinc and lead, and metals such as gold, silver, cadmium and tin are common byproducts in these deposits. There are

also barren massive sulphide bodies that contain only massive pyrite or pyrrhotite.

It is common for the rocks stratigraphically below the mineralized horizon to have stringer or stockwork mineralization, that is, sulphide minerals in veinlets or disseminated through the rock. This mineralization can contain enough metal to be ore, although the grade is much lower than that of the massive ore in the true stratabound body above it. The surrounding host rocks show very strong hydrothermal alteration and sometimes are converted entirely to chlorite, a green alteration mineral.

It's useful to note that the stockwork mineralization is younger than its host rocks, or epigenetic, while the stratabound body is syngenetic, even though the two formed at

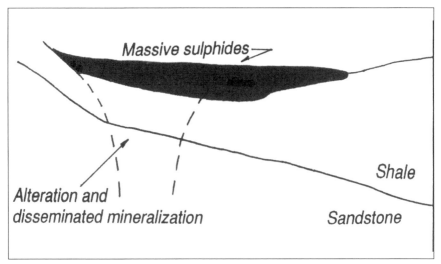

Shale-hosted massive sulphides are usually rich in lead and zinc.

about the same time.

Sedimentary exhalative deposits are similar to the volcanogenic massive sulphides; they are massive sulphide bodies that occur in sedimentary rocks, most often in thick sequences of shale. These deposits also have disseminated or stringer mineralization in the rocks below the mineralized layer and, like their volcanic cousins, may be deformed or faulted into complex shapes.

Some typical volcanogenic massive sulphides are the La Ronde and Kidd Creek deposits of Canada. The Rammelsberg deposit in eastern Germany is typical of the sedimentary exhalative type.

CARBONATE-HOSTED DEPOSITS

Carbonate-hosted deposits are bodies of sphalerite, galena and iron sulphide (pyrite or marcasite) in limestone or dolostone. These deposits are also called Mississippi Valley-type deposits, since that part of North America has many of them.

The deposits are generally believed to be formed when mineral-laden fluids travel through fractures or pore spaces in the rock. Chemical conditions in the carbonate rocks cause the metals to precipitate from the fluids, to be deposited on fractures and in openings in the rock.

The deposits usually have a tabular shape on a large scale, but in detail, because they form in areas where the rock is fractured, broken or caved, they can have an irregular shape.

RED-BED COPPER

Stratiform sedimentary deposits, also known as "red-bed" deposits, are fine-grained disseminations of base metal sulphides, usually copper and iron sulphides, sometimes with lead, zinc, cobalt and silver minerals or native copper. The host rocks are normally shales or sandstones. The minerals appear to have precipitated in the pore spaces of the sedimentary rocks from fluids circulating in the rock. They can occur, in similar form, in layered volcanic rocks.

Geologists' interpretations differ on whether this happened at the time the host rocks were deposited or later on, but both schools of thought agree that the minerals were deposited when the fluids reached a "chemical trap", a place where the chemistry of the rock changed in a way that made it impossible for the metals to remain in solution. The large copper deposits of Zambia and the Democratic Republic of the Congo, are of this type.

There are also sandstone-hosted lead-zinc deposits similar in form to stratiform copper deposits, but usually without iron sulphide minerals.

SEDIMENTARY URANIUM DEPOSITS

Sandstone-hosted uranium deposits and shale-hosted uranium deposits are exactly what their name implies, that is, uranium minerals in a sandstone or shale host rock. The uranium minerals are usually disseminated through the sedimentary rock.

The sandstone-hosted types appear to have been formed in a similar manner to the stratiform copper deposits, with circulating fluids carrying uranium through the rock and then depositing it when they reach a chemical trap. The uranium deposits of Niger and the southwestern United States are sandstone-hosted.

Uranium in these sorts of settings may have occurred through erosion and chemical removal of traces of uranium from other rocks and concentration in the basin where the shales were forming.

Each of the sediment-hosted deposits described above – carbonate lead-zinc, red-bed copper and uranium – can be deformed by later folding and faulting. However, because they form in stable sedimentary basins where there is usually less tectonic activity, they remain undeformed more often than the exhalative types.

PLACER DEPOSITS

Probably the simplest of the syngenetic deposits is the placer. If a mineral is chemically stable and physically resistant, it can be eroded from its primary hard-rock occurrence and transported to river channels, deltas or other sedimentary environments where it can be deposited in a sedimentary bed. Gold, platinum, diamonds and other gemstones are found in placer deposits because they can withstand most geological underground and surface processes, remaining intact after most other mineral constituents of a rock have been altered or oxidized.

The deposit can remain unconsolidated or can be buried and ultimately turned into a sedimentary rock. The gold-bearing conglomerates of South Africa and the uranium-bearing conglomerates of Ontario are widely thought to be old gravel placers that have solidified into rock.

Placer diamonds have been found in India, Brazil, Venezuela, the Democratic Republic of the Congo and several other African countries. In recent years, the mining of placer diamond deposits has even been extended to the ocean floor, notably off the southwestern coast of Africa. But just as all placer gold has its origins in bedrock of one type or another, placers of gem minerals count certain types of bedrock as a source.

COAL, POTASH, SALT AND GYPSUM

These materials occur in bedded deposits in sedimentary rocks. They are not ores; rather, they are extracted and used as is. Coal forms from decayed plant material, hence the name "fossil fuel." Potash, salt and gypsum are left over when seawater evaporates in shallow basins. All these deposits have a tabular form.

LATERITE DEPOSITS

Under tropical conditions, fresh rock weathers very quickly and to great depths. Rocks in tropical climates weather to form laterite, a soft, deeply weathered mixture of oxide and hydroxide minerals and clays. Some metals may be leached away by the weathering process, but others such as aluminum, iron, nickel and cobalt can remain as oxides or silicates. Laterites produce virtually all the world's aluminum ore and lateritic nickel-cobalt deposits are a significant source of those metals.

IRON FORMATIONS

Iron oxides, sulphides, silicates and carbonates often form as chemical sediments on the sea floor. The iron minerals may be of sufficiently high grade and have good enough metallurgical quality to be suitable for steelmaking. The most useful are the oxide iron ores, magnetite and hematite, but iron carbonates may also be shipped as iron ore. As well,

some iron formations host econoic concentrations of gold. Because the deposits are primary sedimentary beds, they are usually tabular in shape.

INTRUSIVE ROCKS AND MINERAL DEPOSITS

Intrusive rocks, particularly granitic rocks, have a hand in forming many different kinds of deposits. Some of the deposits form in the intrusions themselves, while others form in the surrounding country rocks as a direct result of the intrusive activity. This is because with intrusion comes fracturing and hydrothermal activity.

Magmas can intrude into overlying rock gently or forcibly. If they intrude gently, they normally incorporate pieces of the country rock, which melt and become part of the magma. If they intrude forcibly, they can fracture or brecciate the country rock, creating spaces for hydrothermal fluids to circulate.

As an intrusive rock cools, the major rock-forming silicate minerals – like feldspar – crystallize, and the remaining magma is left with the chemical constituents that have the lowest melting point. It contains large amounts of water, carbon dioxide and sulphur, and also has a lot of quartz. Intrusive rocks cool from the outside in, and this fluid-rich "late magma" is generally held inside the intrusion.

The "late magma" can be forced out of the magma chamber into fractures in the intrusion and in the country rock, cooling there and forming veins. Heat from the intrusion can also cause water to

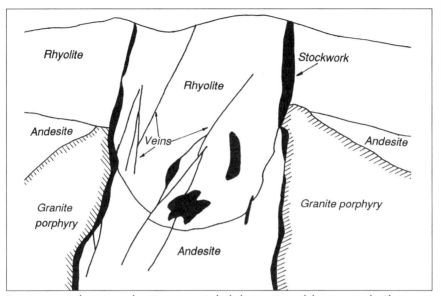

Ore zones in porphyry copper deposits can occur in both the intrusion and the country rocks. The ore can occur in fractures, stockworks and replacement bodies.

be released from the country rock itself, which may begin to circulate through fractures in the rocks and can carry minerals in solution, re-depositing them wherever it finds an open space. The circulating water can also move into porous country rocks such as sandstones or carbonates, leaving behind minerals that have come out of solution as the fluids lost heat or encountered different chemical conditions. And, as always, circulating hydrothermal fluids react with the rocks they pass through, causing alteration.

PORPHYRY DEPOSITS

Porphyry deposits are typical of the mineral deposits formed by igneous activity. They contain any combination of copper, tungsten, molybdenum and gold, but the classic examples are the porphyry copper deposits of the South American Andes. Porphyry-type deposits are large, and the grade of mineralization is usually low.

Both the intrusion and the country rocks are densely fractured, so the mineralization typically forms veins or breccia bodies in the intrusion itself or in the country rock around it. The fractures are often radial, extending out from the intrusion like the spokes of a wheel, or concentric, with the intrusion at the centre. The veins are usually made up of sulphide minerals, plus quartz, calcite, dolomite and fluorite gangue. The metals occur in minerals like chalcopyrite, molybdenite, wolframite and native gold.

The rocks around the deposit are highly altered since large volumes of hydrothermal fluids have circulated through them. In the rocks farthest away from the intrusion, many of the primary minerals are leached. In rocks closer to the intrusion, the primary minerals are completely destroyed and replaced by micas and clays. At the centre of the hydrothermal system, where it is hottest, hydrothermal feldspars are formed.

Another process, while it affects many different kinds of deposits, is particularly important to porphyry coppers. The process is called supergene enrichment, and occurs after the primary copper deposit has formed. Weathering processes leach copper out of the upper parts of the deposit; the copper-bearing solution travels downward, through fractures and pore spaces, until it meets the water table. Because the copper is no longer in contact with the air, it becomes unstable in solution. It precipitates, forming copper-rich minerals like bornite, chalcocite and cuprite, even native copper. This enriched zone can have substantially higher ore grades than the primary mineralization.

Skarn Deposits

Other kinds of deposits are related to intrusions. Skarns form at the contact between an intrusive rock and a carbonate rock or a clastic sediment rich in carbonate. They are zones with irregular shape, and have a characteristic mineral composition – calcium, magnesium and iron silicates. Skarns may contain iron, gold or zinc, but are particularly important as hosts of copper, molybdenum, tungsten or tin ores. They frequently occur on the margins of the intrusions that create porphyry copper deposits.

Granite-Hosted Deposits

Deposits of tin, tungsten, molybdenum or uranium commonly occur in and around granites, either in veins or very coarse granite dykes, called pegmatites. These coarse-grained, granitic bodies are a source of many rare minerals, several of which can be cut into gems. Pegmatites are commonly found as dykes in a large mass of plutonic rock of finer grain size. These deposits can also be useful sources of fluorite or gemstones. The tin deposits of southwestern Britain are an example of these deposits.

Besides pegmatites, minable quantities of emeralds, rubies and sapphires have been found in crystalline limestones, mica schists, syenites and other rocks.

Epithermal Veins

Epithermal vein deposits are large vein systems, usually in volcanic rocks. However, the deposits, the volcanic rocks that hold them, and the intrusive rocks that are always beneath them are the product of a single process. The volcanic rocks are extruded at the surface, but the volcanic vents have feeder stocks and dykes that ultimately cool to form intrusive rocks. The vent areas are centres of intense fracturing and hydrothermal activity, just like the areas around an intrusion.

The mineralized zones are usually veins in radial or concentric fractures, but it is also common for the mineralization to extend into the country rocks as disseminations and replacement bodies. The veins normally contain base metal sulphides like chalcopyrite, galena and sphalerite, but they are most important as sources of gold and silver, which occur as native elements, or in compounds or as impurities in other minerals. The veins also have worthless gangue material, like quartz, fluorite, calcite or dolomite.

Sometimes the narrow veins are mined as high-grade, low-tonnage deposits, and are often called "bonanza" veins for these high grades. In other places, the veins and the country rocks are mined together in bulk in large, low-grade operations.

REPLACEMENT DEPOSITS

Replacement deposits are bodies of rock that have been "replaced" by ore. To be more precise, the ore-forming fluids have migrated out through the porous rock (usually, though not always, a sedimentary rock) and left behind enough mineralization in disseminations, fracture fillings, and true replacements to make ore. Vast low-grade gold deposits in the southwestern United States, such as Carlin in Nevada, are an excellent example of this kind of deposit.

Replacement deposits are often affected by supergene processes. The primary, unoxidized ores frequently have gold bound in pyrite, which prevents many metallurgical processes from extracting the gold.

The weathered ores, in which the pyrite has been destroyed by oxidation, have free gold that is extracted more easily.

LODE DEPOSITS

Lode deposits are a very important source of precious metals, although they also can contain base metals. They are the dominant gold deposits of the Precambrian shields of Canada, Australia, South America and Africa – veins and shear zones, holding native gold, pyrite, quartz and carbonate minerals. They can also form stockworks or dissemination zones, or occupy "saddles" in the hinges of folded strata.

Deposits like these are found in greenstone belts, areas of metamorphosed volcanic and sedimentary rocks. Greenstone belts are most numerous in Precambrian shields, but occur throughout geological time. The deposits themselves can have almost any host rock, but mafic volcanic rocks, felsic intrusive rocks and some sedimentary rocks are the most common ones. Massive sulphide bodies and iron formations can also be favourable host rocks. More often than not, they occupy the contact between two different rock types.

One characteristic shared by all the lode deposits is their occurrence in tectonically deformed zones. The deposits cluster around large regional fault zones, and the mineralized bodies themselves are in zones of intense structural deformation, with fracturing, fault brecciation and shear zones.

Intense deformation makes for a wide variety of shapes and forms for the deposits. Shear-hosted gold deposits can be comparatively straight; shear zones have been very accurately described as "zones of straightening" in which all the planar features in an area – beds, fractures, faults and dykes – are squeezed into near-paral-

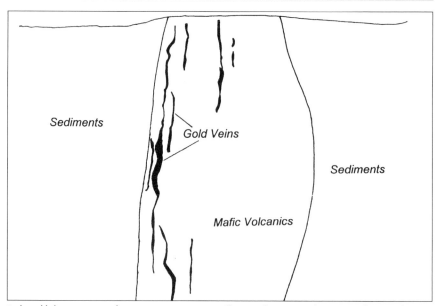

Lode gold deposits most often occur in veins, especially near the contacts between rock units. They are usually narrow but can persist to great depths.

lelism. Veins can be straight or quite sinuous, disseminated mineralization can have almost any form, and stockworks and saddles often form plunging, crudely pipe-like zones. In turn, any of these shapes can be folded into more complex ones, or be cut off by a fault and reappear somewhere else.

The wallrocks of the veins often show very strong alteration. The mineralization in the veins themselves can be very irregular, making it necessary to drill and sample these deposits very closely before making any attempt to calculate their grade and tonnage.

NICKEL-COBALT VEIN DEPOSITS

These are vein deposits that are formed in open fractures. They can be in any number of different host rocks, although they are most common in crystalline metamorphic rocks, granitic intrusions and sediments. Such deposits are rare in volcanic rocks. They contain nickel and cobalt arsenides and may contain native silver, argentite, pitchblende, pyrite, chalcopyrite, sphalerite and galena; the gangue is any or all of quartz, calcite and dolomite.

Examples include the Cobalt district of Ontario, Freiberg in eastern Germany and Jachymov in the Czech Republic.

Without previous structural disturbance to cause the fracturing, the deposits would have nowhere to form. However, the deposits themselves are almost always undeformed. This suggests that they are emplaced very late in a region's geological history.

Most geologists who have studied these deposits recognize that the veins are hydrothermal, but no single genetic model is accepted widely.

UNCONFORMITY URANIUM DEPOSITS

The unconformity-type deposits are the world's main source of uranium.

These deposits form at or near the contact between an overlying sandstone and underlying metamorphic rocks, often metamorphosed shales. The orebodies have lens- or pod-like shapes, and most often occur along fractures in sandstone or in basement rocks. The host rocks often have disseminated uranium minerals and show hydrothermal alteration, which may indicate that the deposits formed after the rocks. The mineralized bodies may carry minor amounts of sulphide minerals like pyrite, arsenopyrite, galena and sphalerite, as well as nickel-cobalt arsenides.

Because this type of deposit is a relatively recent discovery – the first huge uranium deposits in Saskatchewan and northern Australia were found in the late 1960s and early 1970s – geologists are still trading theories about their origin.

A model that has gained favour in recent years suggests that fluids with dissolved uranium and other metals, moving through the sandstone, encountered the basement rocks, where chemical conditions were ideal to cause the metals to precipitate from solution.

Iron Oxide-Copper-Gold Deposits

This is a large and significant type of mineral deposit, but among the most poorly understood. They generally occur in felsic volcanic and intrusive rocks but may take a number of different forms. They can be tabular bodies, or large brecciated zones; they can also occur as veins and disseminations. They may also form replacement bodies in the country rocks surrounding the intrusions.

Their essential characteristic is an abundance of iron oxide, either

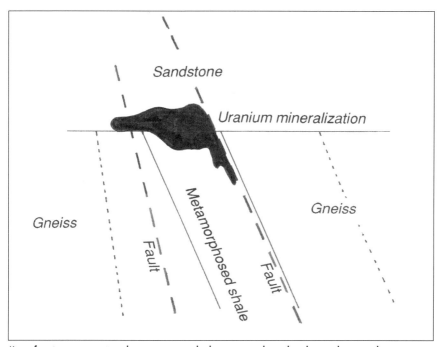

Unconformity-type uranium deposits occur in the basement rocks and in the overlying sandstone.

magnetite or hematite. As such, they can be sources of iron ore, like the giant Kiruna deposit of northern Sweden — but they can also contain large concentrations of other minerals. One such orebody is the Olympic Dam deposit in South Australia, which is a major producer of both copper and uranium. Others, for example the Sossego and Salobo deposits in Brazil, contain copper and gold. Silver and rare-earth elements are also found in some of these deposits.

A significant characteristic is the deposits' regional geological setting: they almost invariably occur in areas of rifting or crustal extension, and in association with large volumes of felsic volcanic rocks.

The mineralization forms through hydrothermal processes: the wall rocks are usually (but not always) altered, with hematite, calcite, epidote, albite and chlorite the commonest alteration minerals. The source of the fluids — whether they come from the cooling igneous rocks or from the surrounding rocks, or are a mixture of the two — is one of the liveliest areas of debate about the origin of these deposits.

4. High-Tech Prospecting

"Boot-and-hammer" prospecting has always been an important part of mineral exploration in Canada. But exploration has changed dramatically with recent advances in technology, all of which play a part in the task of prospecting.

Typically, geological reconnaissance begins the process. In many countries, government geological surveys employ mapping geologists, who examine large areas, making note of all pertinent geological features as revealed in outcrops and prominent landforms. The geological reports and maps they make serve as an important source of reference for the mine finder. Experienced prospectors plan their search in areas where the rocks and geological structures suggest there could be mineralization.

The first order of business for the prospector is geological mapping, on a scale more detailed than that of the government geologists, and surface prospecting.

The prospector looks for trace amounts of ore minerals, for favourable rock types, and for alteration that may have been caused by mineralizing solutions. One valuable sign of mineralization is a gossan, an area of rusty staining on rock formed when sulphide minerals are oxidized. Other ore minerals also can oxidize, leaving a surface stain of secondary minerals on host rocks. The light green of nickel bloom or the bright yellow and orange of secondary uranium minerals are examples of this.

If a showing is found, it is sampled, and the samples sent for a chemical analysis called an assay. Sampling, which will be covered in more detail in Chapter 5, can be as simple as banging off a piece of rock from an outcrop. Often the exploration crew will bring in a bulldozer to strip away overburden or use explosives to blast a trench in the rock.

Another useful technique at the reconnaissance stage is remote sensing, the use of photographic and radar images taken by aircraft or satellites. Aerial and satellite imagery can show large-scale geological structures like faults or geological contacts in which mineralization often occurs. In some areas, such as deserts, colour changes on satellite imagery may denote changes in rock type or show areas of rock alteration.

It is generally believed that tomorrow's mineral discoveries in established mining areas will likely come at greater depths than today's known orebodies. They might also come from areas covered by heavy blankets of overburden. Finding these deposits requires more sophisticated technology than traditional prospecting methods. Buried targets can be explored by diamond drilling, but to know where to aim the drill the prospector will have to use geological inference, geophysics or geochemistry.

We'll examine geophysics and geochemistry in this chapter.

GEOPHYSICS

The rapid expansion of technological knowledge following the Second World War has permitted great advances in geophysics, the study of the physical properties of the Earth. It's not a new science – as early as the seventeenth century, Swedish prospectors were using magnetized iron bars to locate magnetic bodies of iron ore – but it developed rapidly during the postwar boom. And computerization has meant the amount of data we can acquire and process on site has increased dramatically.

In all geophysical surveys, what is sought is an anomaly – an exception to the norm. A geophysical anomaly is an area where the earth has unusual physical properties.

In a typical geophysical survey, a physical property like the gravitational or magnetic field is measured on a grid of locations over the survey area. The value found at each grid position is plotted on a plan view of the property. Lines are then drawn through points having equal value, in exactly the same manner that isobars are drawn on a weather map or elevation contours on a relief map. This map of the geophysical data allows the prospector to pick out areas with the geophysical characteristics that suggest there may be mineralization.

Another approach is to compile the data as profiles, giving a section view of the anomaly. These are commonly used by both geophysicists and geologists, who use them as guides in selecting the most appropriate means of further investigating the anomaly.

Computer-generated models of the kind of geophysical anomaly a particular body might cause are often calculated. For example, it is possible to predict the shape of a magnetic anomaly over a body of regular shape and known magnetic susceptibility. A "forward solution" of this kind helps narrow down the likely physical explanation for anomalies found in actual surveys.

MAGNETIC METHODS

In magnetic surveying, the geophysicist measures the strength of the earth's magnetic field, which will vary locally depending on the amount of magnetic material in the underlying rocks. (Magnetic surveying has also been used, with considerable success, to find buried metal objects like underground fuel tanks.)

Where the rocks have high magnetic susceptibility, the local magnetic field will be strong; where they have low magnetic susceptibility, it will be weaker. This has two applications. Firstly, deposits with magnetic minerals – iron deposits, pyrrhotite-bearing nickel deposits and skarns – can be detected directly using magnetic surveying. Secondly, magnetic surveying can be used as an aid to geological mapping. Units with higher susceptibility will show up as areas of high magnetic field strength.

Magnetic surveys need not be done on the ground. An aerial magnetometer is an extremely sensitive instrument which is either trailed below an airplane or a helicopter, or mounted on the aircraft in a so-called "stinger." By combining readings from this instrument with continuous aerial video photography, a magnetic map of a large area can be plotted. Government geological agencies frequently contract for aerial surveys, publishing the results in order to encourage exploration.

In airborne surveys, geophysical instruments are often towed behind aircraft.

RESISTIVITY

In this method, an electric current is generated and forced into the ground from widely spaced electrodes. The current flows through the earth to complete the circuit, and the amount of current that flows depends on the resistance the rock offers. This can be measured by probing the ground with pairs of electrodes connected to sensitive voltmeters.

A conductive orebody containing economic metallic sulphides will cause an anomalously low resistance. So, too, will a fault plane lined with graphitic material, a barren sulphide or a fracture containing brackish solution. Results from this method must be interpreted using geological evidence.

INDUCED POLARIZATION

Induced polarization, or IP, is a phenomenon discovered in the early days of resistivity surveying. It was found that certain bodies were "polarizable" or "chargeable" – that is, they could be caused to take an electric charge by passing a current through them. When the current was switched off, the charge did not disperse all at once, but drained away.

Resistivity and IP are normally conducted as a single survey. An electrical current is sent through the ground and the surfaces of metallic minerals become charged. An over-voltage has to be applied to drive the current across these barriers. When the current is switched off, the over-voltage decays. In other words, there is a brief storage of energy that can be measured even after the current is switched off.

The IP effect is particularly useful in detecting disseminated sulphide minerals, which may be economic in themselves or may serve as pathfinders to other mineral deposits.

Magnetometer surveys measure variations in the Earth's magnetic field caused by magnetic minerals.

SPONTANEOUS POLARIZATION

A conductive body extending both above and below the water table can act as a weak natural electric battery, creating an electric current in the rocks and soils surrounding it. Using a sensitive voltmeter, it is possible to detect the voltage difference along the flow of the current, which can indicate where the conductor is. SP, one of the first electrical techniques to be developed, is used only rarely now, as other methods are generally believed to be more sensitive.

ELECTROMAGNETIC METHODS

Electromagnetic, or EM, methods are a useful and rapid way to detect buried conductive bodies. An alternating current is fed into a wire coil held in a prescribed direction, either parallel or perpendicular to the ground surface. This current produces an alternating magnetic field, which induces a current in any nearby electrical conductors. Any induced current creates its own alternating magnetic field, which is measured by a search coil connected to a sensitive voltage meter.

The method detects conductive bodies, not mineralization. A conductor could be an economic deposit of metal sulphides, but also a barren pyrite body or a zone of conductive graphite.

Traditional EM methods had little ability to "see" more than about 100 metres below surface. Newer, low-frequency EM methods, "pulse" techniques such as UTEM, and magnetotelluric methods that use the earth's own EM field have increased the depth penetration of

EM prospecting. Another rapid and inexpensive EM technique is the very-low-frequency (VLF) method, which uses the signals from marine-navigation radio stations as a primary field source.

Because EM surveys do not require electrical contact with the ground, they are among the most useful techniques in airborne geophysics. Usually an airborne EM survey is followed up by ground EM work. Electromagnetic surveys can also be done using probes lowered down drill holes. These down-hole surveys are used at more advanced stages of exploration, where some drilling has already been done.

A related method to electromagnetics, with similar field practices, is ground-penetrating radar (GPR). Ultra-high-frequency radio waves are transmitted into the ground from the surface, and are reflected back by underground layers. GPR is useful for shallow surveys indicating depth to bedrock, and for environmental work, as well as for reconnaissance prospecting.

Telluric Current Methods

Naturally-occurring electrical discharges in the atmosphere, called telluric currents, can also be a source of electromagnetic waves for geophysical surveying. A lightning strike, for example, is an electromagnetic pulse that will cause transient currents in conductive bodies in the earth; so are electrical currents in the Earth's ionosphere.

Measuring the voltage differences caused by these currents between stations on the surface allows sub-surface conductivity to be mapped; often telluric-current techniques can map these variations at very great depths.

"Controlled-source" telluric methods, where geophysicists use generators and current transmitters to create telluric effects artificially, are also in use; because the currents are man-made and therefore more predictable, the survey results can be interpreted more easily.

The Gravity Method

The force of gravity is not uniform over the whole surface of the earth; it is actually slightly stronger where the underlying rocks are more dense and slightly weaker where they are less dense. The difference is tiny, but can be measured and mapped, giving the geophysicist another weapon in his arsenal.

Gravity surveys, whether carried out on the ground or from a helicopter, use extremely sensitive balances to detect the variations in density of the underlying rocks. They can be useful in conducting a rapid reconnaissance survey of an area to delineate major rock types. This information can help to indicate areas favorable to exploration by other methods. They can also be used in more detailed exploration to detect mineral deposits, which are commonly denser than the rocks that surround them.

Seismic Methods

For hundreds of years, humankind has had instruments that can measure the amplitude and the direction of shock waves produced by earthquakes. These shock waves are acoustic waves, just like sound waves. And like sound waves, they travel faster in rigid and dense bodies than they do in less rigid and less dense ones. They also reflect from the boundaries between different rock types, allowing the geophysicist to measure the time they take to travel and determine the structure of the rocks below.

Seismic prospecting is the most widespread geophysical method in petroleum exploration. Small artificial shock waves are generated at a selected point by either firing a charge of explosives in a shallow drill hole or dropping a heavy weight. The speed of the shock waves is measured by timing their arrival at sensitive receivers called geophones placed along the survey line.

RADIOMETRIC METHODS

The presence of radioactive elements can be determined by the familiar geiger counter. The instrument measures the energy released during the process of radioactive decay. As a uranium molecule decays, for instance, three kinds of rays are given off: alpha, beta and gamma. Of these, the gamma ray is the most penetrating and is therefore the most likely to be detected by the geiger counter. The geiger counter, because it has a low operating efficiency and a low sensitivity, has in many applications been supplanted by an instrument known as the scintillation counter or scintillometer.

Gamma-ray spectrometers are an even more advanced version of the scintillation counters. They can distinguish between radiation from the three main radioactive elements that occur in nature – uranium, potassium and thorium – by measuring the energy of the radiation. Radiometric surveying can also be done from aircraft.

Ground radiometric surveys are most useful to detect showings of radioactive minerals directly. Airborne radiometric surveys are often used for geological mapping, because the radioactive elements occur in greater abundance in granitic rocks. Down-hole probes are frequently used to measure the radioactivity of rock units encountered in drill holes.

GEOCHEMISTRY

A mineral deposit could be described as an area where certain substances – metals or minerals – exist in concentrations that are much higher than normal. In other words, the deposits themselves are "anomalies." Finding anomalous concentrations of elements is the aim of geochemical exploration.

Exploration geochemists have traditionally distinguished between "primary" and "secondary" distribution of chemical elements. Primary distribution processes are the ones that form the ore deposits in the first place, and disperse metals and other chemical elements through the surrounding bedrock. Secondary processes like weathering, glaciation and the movement of ground and surface water move the elements around more – into soils, stream and lake sediments and waters, out to sea, and even

Electromagnetic surveys are used to detect underground conductive bodies.

into plants and the atmosphere.

A detailed knowledge of the geological processes that cause dispersion is required to trace the metals back to their source. This is the realm of geochemistry. In typical geochemical surveys, the prospector takes samples of a particular material, perhaps bedrock, soil, water or something else entirely. The samples are then analyzed chemically for the elements of interest, and the results plotted on a map.

In the simplest case, the prospector will follow up on samples that have the highest concentrations of the metals he is looking for. More often, the geochemist has to interpret the patterns of distribution to reconstruct the path the elements have followed. For example, glacial action can "smear" soils along the

direction the glacier travels, and the geochemical pattern will be moved away from the mineral deposit that created it.

A relatively new geochemical method is mobile metal ion geochemistry, or MMI. In this technique, soils are analyzed using a very weak chemical extraction process, one that strips out only the most weakly bonded metals. With exceptionally sensitive instrumental analyses, the small concentrations of very mobile metals can be mapped, and the technique has proved useful in many environments and over many different types of mineral deposit.

Geochemistry is used routinely in most exploration programs. In temperate areas that have residual soil, it is frequently the most important exploration technique.

In areas that have been glaciated, the use of geochemistry is often more complex, depending on the exact nature of the glacial history of the area. In the tropics, deep weathering of soils can make geochemical patterns very obscure.

INDICATOR MINERALS

The metals or minerals the prospector wants to find may not be the only materials useful in locating a mineral deposit. Other minerals, not themselves economic, may commonly occur along with ores and may offer a way to prospect for the deposits.

As we mentioned earlier in this book, chemical alteration of the wall rocks around hydrothermal ore deposits may create new minerals, some of which may be guides to mineralization.

In a similar way, metamorphism can cause ore minerals to react with gangue minerals, producing new minerals that are characteristic of metamorphosed mineral deposits. Finding minerals like these in otherwise unmineralized rock should suggest to the explorer that mineralization may not be far away.

Host rocks themselves may provide indicator minerals. An important example of this is the use of indicator minerals to find the kimberlite host rocks of diamond deposits. Diamonds themselves are exceptionally rare, but kimberlites host a number of characteristic minerals incuding pyrope (magnesium-aluminum) garnet and chromian diopside, one of the pyroxenes. These are relatively heavy and resistant minerals, which may occur in residual soils over kimberlite pipes, or may be held in the train of material smeared down-ice by a continental glacier. Finding these minerals and tracing them back to their source is a useful technique in prospecting for diamonds.

An analogous technique is boulder-tracing, used frequently in glaciated terrain. Mineralized "float" — cobbles or boulders in the glacial drift — must have come from a bedrock source somewhere. Exploring the area up-ice from the occurrence of the float can lead to a mineral occurrence in the bedrock.

NO BEST METHOD

There is no one best geochemical or geophysical method. Depending on the kind of deposit a prospector is looking for, one or more methods may be very useful, some may provide helpful additional information, and others may have no use at all. Usually, a well-designed exploration program uses several different methods, chosen to fit the geological environment the prospector is exploring. Often, the best methods can be found by trial-and-error surveys over known mineralization in the area, a technique called orientation surveying.

Lightweight geophysical prospecting tools are making the prospector's task less arduous and more precise. But there is still no substitute for geological insight, and no easy road to a mineral discovery.

5. Sampling & Drilling

SAMPLING

Sampling is the process of taking a small representative portion of a larger mass. By analyzing the sample to determine the concentration of metal it contains, the potential value of the larger mass can be determined.

The first samples taken from a mineral showing are called grab samples. Prospectors and geological field crews gather grab samples from outcrops, road cuts, trenches or river beds. These rocks are selected specifically because they appear to contain a significant amount of metal, so they are not considered representative of the outcrop or road cut from which they come.

In the field, grab samples are gathered, their original location is recorded, each rock is labeled and the most promising ones are sent to a lab for metal analysis.

If worthwhile or significant amounts of metal are present in such grab samples, channel sampling may be warranted. In this sampling technique, bedrock where the sample was taken is exposed

as fully as possible, typically by using a backhoe or some such piece of earth-moving equipment.

Next, the outcrop is hosed down with water and, if a zone of mineralization is revealed, representative surface samples are taken at regular intervals across the exposed zone. These samples are usually cut with a portable circular saw equipped with a diamond-studded blade, leaving a linear channel across the outcrop.

The surface channel is the most desirable type of sample. It is normally cut about 10 cm (4 inches) wide and 2 cm (3/4 inch) deep across the supposed ore zone. The chips of rock removed are carefully collected, marked and bagged for analysis.

Chip samples are sometimes taken by the geologist or engineer for a quick approximation of contained value. Random pieces are quickly knocked off the outcrop with a hammer and chisel, with an effort made to take representative amounts. Chip samples cannot be relied upon fully, so they generally do not enter final mathematical calculations of possible reserves.

It is highly desirable, but not often practical, to space surface channels at regular intervals along the mineralized zone. This obviates one mathematical calculation in the interpretation process.

In certain circumstances, particularly when sampling kimberlite rock for diamonds, it is useful to collect a bulk sample, which may range from a few hundred kilograms to several tonnes in weight. It is important for a bulk sample to be representative of the zone since this material can be used later for definitive metallurgical test work and grades.

DIAMOND DRILLING

Just as one should not judge a book by its cover, surface sampling gives no definitive indication of how tremendous - or how mediocre - a deposit lies underfoot. Thus, after surface sampling indicates a possible concentration of valuable mineral, diamond drilling is undertaken.

The only way to ascertain the quantity (tonnage) and concentration (grade) of a deposit is to make a circular cut in the rock and extract the continuous cylindrical core sample from the centre of the cut. To do this, a special type of drill has been developed, with a rotating core barrel that grinds down through the bedrock. At the end of the core barrel is a cylindrical bit studded with the hardest of natural substances - diamonds.

The size of diamond drill core varies with the size of the machine used, hole depths and material being drilled. However, the most common sizes are:

A - core diameter 27.0 mm, hole diameter 48.0 mm;

B - core diameter 36.5 mm, hole diameter 60.0 mm;

N - core diameter 47.6 mm, hole diameter 75.5 mm; and

H - core diameter 63.5 mm, hole diameter 96.0 mm.

Mechanically, the diamond drill consists of a power unit rotating a tubular steel bit on the face of which are set diamonds. This bit and attached core barrel are rotated under controlled pressure by means of hollow steel rods. Water is pumped through these rods to cool the bit and remove the rock cuttings.

Under ideal drilling conditions, and once the drill bit is lowered to the bottom of the hole and the drill started, the bit will cut a core, consisting of a cylindrical piece of rock. Rotating at a high rate of speed, the bit is forced downward by the action of hydraulic cylinders on the drill. As it moves through the rock, it pushes the core up into the core barrel.

The rods are withdrawn at intervals of 1.5 or 3 metres (5 or 10 feet) and the core is removed from the core barrel for examination and storage. The core presents a tangible and accurate record of the various rock formations through which the bit has passed.

Sludge samples are also sometimes taken while drilling is under way. These consist of the cuttings made by the drill, and are useful as checks against the drill core samples.

WIRELINE DRILLING

Most drilling today is done using the wireline method, which was introduced in the 1960s. Using this method, an inner tube containing the

Drilling for gold in northern Ontario.   Photo by The Northern Miner

core is detached from the core barrel assembly when the core barrel is full or a blockage occurs. The tube and core contained in it are pulled to the surface by a wire dropped down the string of drill rods. A latch or "overshot assembly", which snaps on to the top of the inner tube, is used for this purpose. The inner tube is then rapidly hoisted to surface within the string of drill rods.

After the core is removed, the inner tube is dropped down into the outer core barrel and drilling resumes. Thus, the core is retrieved without having to pull all of the rods.

Large drills have recovered cores of up to 100 mm in diameter from depths of more than 4,500 metres (15,000 feet).

Core is carefully placed in sequential order in boxes or trays and taken to a geological field camp or office for examination by geologists. Interesting sections are split along their length by a core-splitter, with half of the section being returned to its place in the core box and the other sent to the assayer for analysis. The trays or boxes are usually stored on racks in a core shack.

Sometimes it is desirable to obtain a second intersection of a particular geological structure or ore zone from a single hole. This is done by placing a wedge at some point above the intersection to deflect the bit in another direction. This wedging procedure is used frequently in deeper holes, where it can save considerable time and money.

OTHER DRILLING METHODS

The reverse circulation drilling process is being used successfully in areas where rocks are deeply weathered, as in many tropical locales, or when drilling in glacial overburden. In this method, a tri-cone bit and dual-tube drill pipe are employed. Drilling fluid or air, or a combination of the two, is pumped down between the dual tubing and returned up the inner tube, bringing

A field crew removes drill core from a core barrel for examination.   Photo by The Northern Miner

cuttings from the bit to surface.

Sonic drilling, in which a sonic vibration device pushes the drill rod into the ground, can be used for soil investigation.

Sometimes explorers don't need a core sample, as when an open hole allows instruments to be lowered. A standard rotary drill equipped with necessary compressors and drill pipe is all that is needed for this type of drilling.

Where applicable, cost reduction is significant together with greatly improved productivity. However, this method is not as reliable as collecting the full core by diamond drilling.

Underground Drilling

Diamond drilling is also important to an established mine. It is used to:
• explore for new ore or outline and map known orebodies;
• investigate rock types and their structure;
• locate orebodies displaced by faults and folds;
• put out pilot holes to direct drifts and stopes to the proper location;

- establish drain holes, grouting and ventilation holes; and
- drill undisturbed holes to allow placement of rock mechanics instrumentation for measuring stresses in the rock.

Underground diamond drills are smaller and lighter than surface drills, and normally run on electricity, not diesel fuel.

Typical modern underground drills have features which make the extraction of rods from the hole very fast.

While most underground drills are now hydraulic types powered by electric motors, some sizes are also available with air and diesel power for when electricity is not available. Diesel-powered drills, however, need extra ventilation and diesel exhaust scrubbers.

Having a drill underground permits the drilling of holes at any angle, which is very advantageous to the mine geologist.

Assaying

The most promising hand samples and core invariably wind up in a laboratory to be analyzed. This process, in which the precise constituents of the rock can be measured and catalogued, is called assaying.

The chemist chooses an assay method that best determines the concentration of the metal of interest. Among the methods commonly used today are:

- fire assaying, in which the sample is melted and the unwanted elements are chemically removed;
- wet assaying, in which the sample is dissolved and metals are recovered chemically using reagents; and
- instrumental analysis, in which the

metals' atomic properties are detected – for example, their response to x-rays or visible light.

After surface assay results are returned from the lab, the location of the samples and their corresponding assay values are plotted on a map to give a two-dimensional picture of the potential ore zone.

In later stages of exploration, core assays are similarly plotted on maps, adding a third dimension that enables the geologist to visualize the entire orebody.

Often in gold exploration, any unusually high values are cut; that is, they are not included in the average. Isolated high values are frequently found not to reflect the grade around the sample location.

If the assay result is from a channel sample or a length of core, it is written in terms of metal concentration over a given length – for example, 5 grams gold per tonne (0.15 oz. per ton) over 8 metres, or 10 feet of 3% nickel. While these are only two-dimensional snapshots of a mineralized zone, enough of them in one vicinity combine to form a three-dimensional picture of the tonnage and grade of a deposit.

Resource Calculations

The quantity, or tonnage, of mineralized material in a deposit can be calculated if the volume of the deposit is known. The volume can be measured by the widths and depths of the drill holes, and the distance between them – it is now usual to make such calculations using sophisticated computer software that also allows a three-dimensional view of the deposit on screen.

Multiplying the volume by the average density of the mineralization gives the number of tonnes in the deposit. Although this sounds simple, orebodies can be quite complex; densities change, grades are often discontinuous and the shape of an orebody is often highly irregular.

The grades determined by sampling the drill cores are plotted, and a weighted average of the grades is calculated, giving a resource figure. This estimate gives the grade and tonnage of the deposit, so far as it is known by drilling. More tonnage can be added by drilling areas that have not yet been drilled, but extend along the strike and down the dip of the mineralized zones.

Resource estimates are classified as "measured" if the drill holes are closely spaced and the geologist is satisfied that the tonnage is reasonably certain; "indicated" if a significant amount of drilling has been done, but some presumed mineralized zones are not fully tested; and "inferred" if the estimate is based on information from widely-spaced drill holes.

Reserve Calculations

While a resource estimate gives some idea of the amount of mineralized material in the ground, it does not imply anything specific about the practical question of mining the deposit. A reserve estimate, on the other hand, refines the resource estimate by placing economic constraints on the size and grade of the material that is brought into the calculation.

The cutoff grade is the grade below which the rock is assumed to be uneconomic to mine. This grade will vary according to such factors as mining costs, metallurgical recoveries, and possible credits from associated minerals that can be recovered as byproducts. In a reserve calculation, any parts of the mineralized zone that have grades lower than the cutoff are not counted into the estimate.

If a mining professional is calculating a reserve estimate for a possible underground mine, it will be essential to include only the mineralized zones that are wide enough to mine. Small zones of waste between and along the sides of mineralized zones will be included in the estimate as dilutive material.

If the estimate is being made for an open-pit mine, a pit must be designed to show the limits of the mineralization, plus the waste rock that must be removed. The stripping ratio is the mass of waste rock that must be removed to mine a unit mass of ore – for example, if a 20-million-tonne orebody is mined from a pit with a total of 80 million tonnes of rock, the 60 million tonnes of waste give the pit a stripping ratio of 3 to 1.

It is now common to make several reserve calculations, each one based on a different cutoff grade and showing different mine or pit designs. The company can then choose from a number of different possibilities when it develops a final feasibility study on the project.

Like resource estimates, reserve estimates are labelled to show how reliable they are – proven, probable and possible. Only proven and probable reserves are considered in a feasibility study.

6. Mining Methods

THE METHOD IS EVERYTHING

There are as many methods of extraction as there are sizes and shapes of orebodies. The shape and orientation of an orebody, the strength of the ore and surrounding rock and the manner in which the minerals are distributed are different for each ore zone, and influence the selection of a mining method and the plan to develop the orebody.

Operating mines range in size from small underground operations (some of which may produce under 100 tonnes of ore per day) to large open pits moving tens of thousands of tonnes of ore and waste rock per day.

GETTING DOWN THERE

The primary opening into an underground mine can be a shaft, a decline (also called a ramp) driven down into the earth, or an adit, a horizontal opening driven into the side of a hill or mountain. All have the same purpose: to provide access for people, materials and equipment and to provide a way for ore to be brought to surface.

Shafts are usually vertical, although they can be inclined, and are equipped with hoists and headframes, structures at the top that enclose the hoist.

Ramps, on the other hand, usually spiral downward at a gradient of about 15% to allow access into the mine by rubber-tired mobile equipment. In some cases, ramps are driven in a straight line to accommodate conveyor belts, or have straight runs with switch back points. Ramps are generally less expensive to develop than shafts. But depending on the angle of the decline, the size of the opening and the ground conditions encountered, the total cost may be higher than the cost of developing a shaft to reach the same depth.

A typical underground mine will have both types of access; usually a main shaft provides routine transport and hoists ore, while the ramp lets equipment drive from surface down to mining levels and provides emergency access or escape.

Horizontal or level mine workings are called crosscuts and drifts. Sometimes it is useful to open vertical workings between levels in an underground mine; these are called raises or winzes. And it is usual to construct raises to the surface to bring in fresh air.

EXPLORATION DEVELOPMENT

Surface drilling will indicate whether a mineralized zone has the potential to become an orebody. To outline the zone with greater accuracy, and confirm that the mineralization is continuous and the estimates of grade and tonnage are correct, surface work is followed by underground development and detailed definition drilling from the workings. Only then can the developer make plans for production.

Diamond drill holes from surface can only tell part of the story. A mineralized zone may pinch or swell or be irregular, and a large amount of diamond drilling is needed to fill in the information between the first holes the developer drills. Sooner or later, drilling long holes from surface becomes prohibitively expensive, and the mine developer must decide whether to continue exploration from underground. If exploration goes ahead, a suitably sized exploration shaft or an access ramp is driven, allowing crews to get closer to the orebody.

The crews drive drifts and crosscuts from the shaft or ramp, and drill stations are excavated at regular intervals to accommodate underground drills. The mineralized body is then drilled from those stations.

Up to this point in the exploration program, the only samples taken from the orebody have been the drill cores. Once underground, however, it is possible to mine a much larger bulk sample of the ore. The mineralized material in this bulk sample can be sent for metallurgical testing, first at "bench scale" in a laboratory and ultimately at "pilot scale" in an actual mill. This allows metallurgists to test their design in a working model before the developer has to build one.

OPEN-PIT MINING

An open-pit mine is the safest and least expensive kind, and is every developer's first choice where an orebody is close to surface, is big enough, and has little overburden.

Open-pit mines look simple, but every pit needs to be tailor-made. First and foremost, the pit walls have to stay up, so a rock-mechanics engineer has to determine a safe slope for the pit. There is also a delicate balance between how much waste rock can be mined in order to gain access to the valuable ore and how deep a pit can be.

The size and location of the first bench of any open-pit mine is critical. It is excavated well into the waste rock surrounding an orebody. And since each successive bench is smaller than the last one taken, the depth to which the pit can be mined is determined by the size and location of the first cut or bench.

The amount of waste rock mined relative to the amount of ore mined is called the stripping ratio. In most cases, this ratio is high for the first

Open-pit mining is favoured for large, near-surface bodies.   Photo by The Northern Miner

bench and decreases steadily with each successive bench. A stripping ratio of 3 to 1 means that during the life of the pit, there will be three times as much waste rock mined as ore. To be profitable, an open-pit mine must be designed so that the cost of mining the waste rock does not exceed the value of the ore.

The main cost advantage of open-pit mining is that the miners can use larger and more powerful shovels and trucks: the equipment is not restricted by the size of the opening it must work in. This allows faster production, and the lower cost also permits lower grades of ore to be mined.

If an orebody is large, and extends from surface to great depth, it is common to start mining near the surface from an open pit. This provides some early revenue while preparations are made for underground mining of the deeper parts of the orebody. It is not uncommon for ore below the floor of an open pit to be developed from underground by driving a ramp from the lower part of the pit.

UNDERGROUND METHODS

Generally, orebodies are either vein-type, massive or tabular in shape. This, together with ore thickness and regularity, will influence the mining method selected.

Vein-type orebodies usually dip steeply, allowing ore to fall to a lower mining level where it can

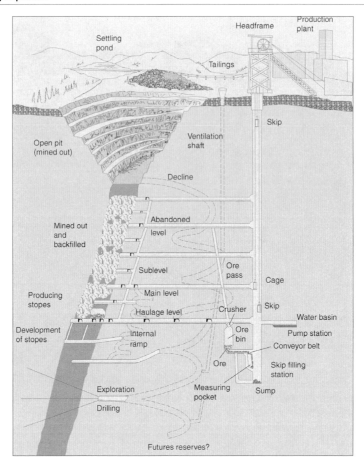

A cross-section of a typical underground mine, showing a shaft, a decline from the base of a surface pit, and a series of main levels. (Drawing courtesy Atlas Copco, www.atlascopco.com.)

be loaded. The orebodies are usually narrow and often irregular, so care must be taken to avoid mining barren wall rock. They are most successfully mined by small-scale underground stoping.

Massive orebodies are large and usually have an irregular shape. Underground bulk mining methods, with large stopes, are best suited to this type of orebody.

Tabular orebodies are flat or gently dipping and the ore, having nowhere to fall, must be handled where it is blasted. Room-and-pillar mining is normally used to extract the ore. Depending on the thickness and lateral extent of the ore, these types of deposits tend to be moderate- to high-tonnage producers.

The strength of the ore and the rocks surrounding an orebody also influence the method (and therefore the costs) of mining the orebody. Openings may be supported or self-supported. Some supported openings are held up by backfill, waste rock or aggregate placed in the openings shortly after they are mined out.

In the old days, miners often supported workings with "sets" made of timber or steel, although this kind of mining is costly and little used today. To support open workings in modern mines, it is more usual to insert steel rock bolts, to bolt up a network of steel straps, or to apply quick-setting concrete to the back and sides of the opening.

If walls and pillars are of sufficient strength to carry both the weight of rock above them and the horizontal stresses in the rock caused by tectonic forces, workings can be self-supporting, although the miner may help them along with rock bolts and screens. In massive orebodies, it is common to plan for the mining of pillars. This is done by backfilling the mined-out stopes to provide the necessary support when the pillars are mined.

The Mining Cycle

The miner's job is to break ore and rock and get it from underground to surface.

The first step in the mining cycle is to drill holes in the rock and ore. These holes are loaded with explosives which are detonated to break the rock. In modern-day mining, electric-hydraulic drill jumbos drill the pattern of short holes required to break the rock or round and advance the development heading. A drift round or breast round will normally advance the drift face about 3 to 4 m (10 to 13 feet).

Older pneumatic jumbos, and hand-held jack leg drills and stopers, are still used for certain specialized drilling jobs and in stopes where the work space is confined.

Large-diameter (10 to 15 cm) blasthole drills, and two-boom, small-diameter (5 to 10 cm) longhole drills are in common use for production in stopes.

Loading and Blasting

Once the holes have been drilled, a blasting crew loads the holes with explosives or blasting agents. A mixture of fertilizer (ammonium nitrate) and fuel oil, called ANFO (an acronym for ammonium nitrate and fuel oil), is the most widely used blasting agent in mines. It is set off by detonating a small amount of high explosive.

Other common explosives, normally used in wet conditions that make ANFO less effective, include water slurries of ammonium nitrate with an explosive like trinitrotoluene (TNT), and emulsion explosives, which combine an oxidizing agent and a fuel. Because they are in liquid form, emulsions and slurries are easy to transport and load and have largely replaced more dangerous dynamite explosives in modern mines.

Machines for loading explosives range from small portable ANFO loaders to specially designed mobile vehicles with large hoppers for transporting and loading bulk explosives pneumatically.

Holes are loaded so that each charge is fired in a designed sequence, often timed in thousandths of a second, and in a pattern that will cause the rock to break in a manner that is safe and convenient for later mucking.

Large blasts or charges are usually set off electrically from surface once all underground workers are out of the mine, usually at the end of a work shift. This is done not only to ensure safety but also to allow the dust and fumes caused by a blast to be dissipated by the ventilation system before the next shift goes underground.

MUCKING

Once the ore or rock has been broken by explosives it is called muck; this ore must be loaded into either trucks or rail cars to be transported to an ore pass for further transfer to surface. A wide variety of loaders and haulers is available to mining companies.

The load-haul-dump (LHD) machine is perhaps the most common mucking machine. These rubber-tired machines, with electric or diesel power, range in size from very small, compact units with buckets that can handle 0.4 cubic metres (14 cubic feet) of muck to the much larger units equipped with 6-cubic-metre (212-cubic-foot) buckets or greater. In mines where haulage distances are great, LHDs are used to load haulage trucks or rail cars, which then haul the ore (or rock) to surface via a ramp, or to an ore pass, a vertical working the ore can be dumped down, like laundry or garbage down a chute. From the ore pass the ore either goes to an underground crusher or is hoisted to surface from a loading pocket near the base of the shaft.

Continuous mucking machines lift the muck from the floor directly on to a conveyor belt which then loads it into a truck or on to another convey-

or. High-speed electric trolley trucks can also be used to bring the ore to surface. Conveyor belts, too, are finding more applications in underground mines (for transporting ore to an ore pass or crusher). Such large-volume mucking systems are usually restricted to mines that use bulk mining methods in the thousands-of-tonnes-per-day range.

BACKFILLING

Once a stope has been mined out, it is often necessary to backfill it with some waste material so that the ore adjacent to the stope can be mined without affecting the structural integrity of the underground workings. Backfill can be waste rock from underground, sand brought down from surface, or mill tailings that have been processed so that all the fine fraction has been removed. This material is often mixed with a bonding agent like cement so that it stands up once the ore adjacent to it is mined out. Backfill may be placed immediately during the mining cycle, or it may be placed once the stope is completely mined out, depending on ground conditions.

ORE HAULAGE

Mines formerly needed great lengths of track for ore haulage, but rubber-tired mining machinery has given operators a choice between trains and trucks. In general, longer haul distances and narrower workings favour track haulage; larger openings and shorter distances favour low-profile trucks or LHDs. Trackless equipment also allows for more flexibility in underground transport, and,

Mechanized equipment is used to haul ore at many underground mines.   Photo by The Northern Miner

unlike rail cars, it has the capacity to travel on inclined workings.

In track mines, ore is drawn from the chutes on various working levels into ore cars. These are hauled ("trammed") by electric or diesel locomotives to a station where the cars are dumped. The ore falls down a short finger raise into a main ore pass. Ore cars vary in capacity from one to 20 tonnes.

The ore pass is a fairly large raise, usually 2.5 to 3.5 metres (8 to 11 feet) in diameter, where broken ore is dumped. A mine's main ore pass extends from the uppermost to the deepest level in the mine. Short finger raises connect haulage levels to the ore pass. Ore falls down the pass to an underground crusher station. Control chutes are usually established at various intervals in the ore pass system to provide ore storage above the crusher. To prevent oversized chunks of ore from plugging up the ore pass, steel rails are placed in a grid-work pattern, called a grizzly, over the dumping station.

Common practice is to install a large jaw or gyratory crusher in the underground crusher station. This unit crushes the ore, usually to less

Broken ore is loaded onto mucking machines for transport to surface.　Photo courtesy of Teck Resources

than 15 cm (6 inches). The ore then falls into a large chamber or bin below the crusher. From here, it is fed into a loading pocket situated near the shaft bottom.

SKIPPING ORE TO SURFACE

The most common conveyance for carrying the ore to surface is the skip. These are self-dumping buckets and are usually operated in counterbalance in two separate shaft compartments to reduce the amount of power needed for hoisting. The weight of the empty skip descending will compensate in part for the weight of the other skip that is being hoisted.

Skips are of lightweight alloy construction and carry loads of three to greater than 20 tonnes. They are filled from the loading pockets at the base of the ore pass.

In some mines, hoisting by conveyor belts may be less costly, and in mines developed by a decline rather than a shaft, it is usual to tram the ore directly to surface by truck or rail car. Whatever way it gets to surface, the ore goes to the mill for crushing, grinding, and beneficiation.

An orebody that dips can be a long distance from the main shaft at depth. Then an internal shaft, called a winze, may be needed for further development; it has its own underground hoist with skips and cages.

At the bottom of any shaft or winze is a sump. It holds the groundwater that seeps into the mine so that it may be pumped either to surface or to the mine's water circuit.

Shafts are divided into separate compartments: one for the skips; one for the cage; one for the manway (a series of ladders for emergency access); and one for service equipment such as water discharge and supply lines, electric power cables, compressed air, and communication and data cables. It is also common to have one or more raises to surface for ventilation, at a distance from the shaft.

Stoping Methods

Stopes are the production centres of the mine. It is here that the ore is first broken.

The safety of the miner is a big consideration in selecting a suitable mining method. Even then, miners must be constantly on their guard to make their workplace as safe as possible. Loose rock is a constant potential danger. As soon as a miner enters a stope, the ceiling (or "back") is examined carefully, and any loose material is brought down with a scaling bar.

A stope must also be designed so miners can reach their working places, remove the broken ore, and get supplies, tools, explosives and equipment in. It must also be properly and continuously ventilated, so that the air does not fill with dust, methane or machine exhausts. In very deep mines, where the wall rocks can be very hot, ventilation also helps to keep the air temperature low enough for miners to work.

Stopes are started from the main levels in a mine. The first step of the operation is known as silling: cutting a drift across the top or bottom of the ore to be mined.

To remove ore from a stope in trackless mines, miners construct a system of drawpoints spaced equally along the footwall of the orebody. These are developed so that they break into the ore at the sill level. Broken ore falls down these drawpoints — which are shaped like funnels to facilitate it — to areas where LHD machines can load it.

Smaller-scale stopes may require the miners to build box holes or raises below the ore zone, with stations underneath for small hauling machines to load.

Development work provides access, ore removal, service and supply to a stope. Raises accommodate mine services like water and electric lines; other development may include rib pillars (oriented at right angles to the strike) and sill pillars (oriented parallel to the strike) over the main haulageway.

Open Stoping

The low-cost, bulk method known as "blasthole" or "open" stoping suits large, regularly-shaped, steeply dipping orebodies. The wall rock must be competent: the stope has to be able to stand open without support. It is also a convenient way to excavate large underground openings, such as crusher stations and storage bins.

Typically, a block of ore is prepared by driving sublevels through the orebody at vertical intervals of about 20 metres (66 feet). Then a raise is made between sublevels, and opened across the width of the stope into a slot shape by successive blasts, providing an opening in which to blast the remainder of the ore in the stope.

Blastholes are drilled in a fan-like pattern into the ore across the entire face of the stope. The ore is blasted so that it breaks into the slot raise, falling to the base of the stope where it can be removed by load-haul-dump machines from drawpoints or by rail cars loaded by slushers.

Vertical Crater Retreat

Drills capable of drilling large-diameter (15 cm) holes up to 60 metres (200 feet) in length have allowed a highly efficient stoping method to displace conventional blasthole stoping in some mining situations. Vertical crater retreat (VCR), also known as vertical retreat mining, reduces the cost of mining wide, steeply-dipping orebodies. The stope has a similar shape, but instead of blastholes drilled in fans, large-diameter holes are drilled vertically from a top sill to break through into a bottom sill on the sublevel below. This allows the ore to be broken into the bottom sublevel in successive horizontal slices using the same blasthole for each successive blast.

Only the bottom of each hole is loaded; the blast breaks off a slice of ore from the bottom of the ore block, which falls into the draw-

point level below, where it can be mucked out. Alternatively, remote-controlled LHDs can load directly in the undercut area of the stope, making room in the stope for the next blast. Dilution is controlled by removing just enough ore to create a sufficient void for the following blast.

This mining method is particularly safe, because the miner does not enter the area where the ore is blasted. Drilling and loading are done from the top sill. The method also eliminates the need to support the ground in the stope after each blast.

Room-and-pillar

In cases where the orebody is narrow and flat-lying, as is the case with many coal, potash, salt and Mississippi Valley-type lead/zinc deposits, a mining method known as room-and-pillar is often used. As the name suggests, ore is mined from large voids or rooms, and pillars of ore are left between the rooms to support the overlying strata. The ore pillars are normally left to support the workings when mining is finished, and open workings may be backfilled.

Production stoping works in much the same way as mining a drift or crosscut; usually horizontal holes are drilled and a round is blasted out and mucked mechanically from the face. Modifications of the room-and-pillar method allow mining in moderately dipping orebodies.

Cut-and-Fill Stoping

Cut-and-fill stoping is suited to irregular orebodies with wall rocks that cannot support loads over large stoping heights. Since backfilling adds a step to the mining of each slice of ore, the ore has to have a high enough grade to offset this added cost.

The stope is mined in horizontal slices or cuts, usually upward from its base. Each slice is blasted on to the floor of the stope and the ore is mucked to the stope mill hole, which leads to the chutes in the haulageway below. Mining a slice leaves a space along the entire length and width of the stope.

After the mill hole and manway structures have been extended, the stope is backfilled, usually with concrete to provide a solid floor, leaving enough work space above. The process is repeated until the stope is mined out.

Cut-and-fill is very flexible. Ore production can be carried out in one part of the stope while the other part is being filled. It also allows miners to selectively mine ore and waste with little dilution, and waste can be left behind in the stope as fill. With a competent ore rock and proper support, it is also quite safe.

Shrinkage Stoping

Shrinkage stoping is a flexible mining method for narrow orebodies that need no backfill during stoping. Successive horizontal slices of ore, usually about 3 metres (10 feet) high, are taken along the length of a stope, in a manner similar to cut-and-fill. The ore is removed from the stope through drawpoints at the bottom horizon spaced about every

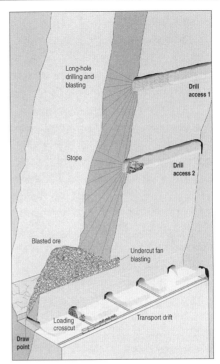

A typical open stope, developed using sublevels. The ore is drilled, then blasted into the open area to the left. It falls to the base of the stope for loading at draw points. (Drawing courtesy Atlas Copco, www.atlascopco.com.)

7.5 metres (25 feet) along strike. Just enough ore is left in the place to provide a floor from which to work when taking the next cut.

Mining continues upward until it reaches the top of the stope, where a horizontal crown pillar supports the walls.

Shrinkage stoping depends on gravity to keep the broken ore moving to the draw points, so it works only in steeply-dipping orebodies. There is no provision for support, so the wall rocks must be strong and competent. The orebody must also be wide enough to allow a working width all the way up the stope, generally no fewer than two metres.

Sublevel Caving

In sublevel caving, ore is developed from a series of sublevels spaced at regular intervals throughout the orebody. Mining begins at the top of the orebody. A series of ring patterns is drilled and blasted from each sublevel. Broken ore is mucked out after each blast and the overlying waste rock caves on top of the broken ore.

This technique is inexpensive, highly mechanized and yields a large amount of muck. It is normally used in massive, steeply-dipping orebodies with considerable strike length. Since dilution and low recoveries are unavoidable, sublevel caving is used to mine low-grade, low-value orebodies.

Stope and Mine Control

The mine has an engineering department that keeps the plans, designs and schedules up to date, and makes sure the ore is extracted safely and efficiently. Surveyors keep a running check on the progress of mine workings and on the volume of ore removed. Safety supervisors ensure that the workings remain stable and that the staff follow safe working practices: or they don't, and sooner or later there is an accident.

The mine geology department keeps up-to-date plans and sections of all levels, including information on ore grades and ground conditions. It also usually directs ongoing underground exploration for more ore. Development headings and stopes have to be sampled: in a large mine, this amounts to hundreds of samples, all of which must be assayed or evaluated on the same day. The results are used as guides and checks on production and serve as the mine's quality control program.

Mine Services

Mines don't just need machinery and explosives. Mines need compressed air, electrical power, ventilation, dewatering or pumping and backfill distribution.

In a day, a typical underground mine handles a greater mass of ventilating air than ore. The deeper the mine, the hotter, so more air has to be moved: the ratio of air to ore can rise to 14-to-1 in the deepest mines. The ventilation engineer maintains up-to-date surveys of the ventilation system to ensure that the workings throughout the mine are kept free of exhaust fumes and dust.

Mines are holes in the ground, and any hole that extends below the water table will eventually fill up with water that seeps in by way of fractures or porous rock units in the walls. Mines are designed so that groundwater runs down to sumps to be pumped out.

Trackless electric-hydraulic and diesel equipment are used more and more often in mines. Increased mechanization allows greater flexibility in underground mining, but it also brings a need for more maintenance. Modern mine designs have to include underground garages, fuelling stations, and repair shops. And there are lunchrooms, wash stations, and refuge rooms for the miners to make the workplace safer and more comfortable.

7. Processing Ore

Winning the Metals

Digging ore from the earth is only half the battle. Often just as challenging and costly is the processing of the ore, which takes place in mills, smelters and refineries.

While the interior of a mill seems, to the visitor, to be a baffling maze of tanks, pipes, pumps, conveyors, motors, chemicals, pulps and solutions, this seeming confusion is actually a carefully designed system constructed for one objective – to recover the valuable minerals locked up in the ore.

The end product from a mill is called a concentrate, or in the case of gold and silver, a doré bar of the metal itself.

All milling and concentrating processes begin with a crushing and grinding stage, which usually represents most of the total cost of processing the ore.

Ore minerals are usually found within and among grains of other ore minerals or (relatively) worthless gangue minerals. This fact makes the processing of some ores more complicated than others.

For instance, a complex sulphide ore containing microscopic particles of sphalerite within small blebs of galena or other sulphides presents a special challenge to the metallurgical engineer, that is, to design a milling process that will liberate these various constituents from each other as cleanly and economically as possible so that each may be recovered.

Primary Crushing

The milling process begins with the primary crusher, which is most often located below the mine workings so that broken ore can be dropped down an ore pass to be crushed and then hauled to the surface in a skip. This is so because loading skips with small, 15-cm-wide pieces of ore is more efficient than loading it with larger chunks.

The primary crusher is usually a

Ore is unloaded at the mill for crushing.

jaw crusher. Ore falls into the opening between a pair of metal jaws at the top and is crushed by the short, rapid, motion of the one movable jaw – a process not unlike an animal using its jaws to chew food.

A few large-tonnage mills use a gyratory crusher as the primary crusher. It consists of a heavy, gyrating head which works inside a crushing bowl fixed to the main frame. Rock falling into the bowl is caught and broken up by the gyrating head.

SECONDARY CRUSHING

A secondary crusher is frequently needed when the product from the primary crusher is too large for efficient grinding. In North America, the main type used is the cone crusher. It is a close cousin of the gyratory crusher, although the speed of the cone crusher is greater and it is designed to handle smaller pieces of rock.

Vibrating screens are used to control the size of the final product from the crusher building. Ore that falls through the openings in a screen is called undersize. It will find its way into the mill on a conveyor belt. Ore that is too large to fall through the openings is called oversize and it returns to the crusher along with a separate conveyor belt. This ore is called the circulating load.

Crusher buildings are typically equipped with dust control and ventilation systems designed to eliminate the buildup of harmful dust.

Most mills operate seven days a week, 24 hours a day, compared with the mine, which may operate only five days a week. If there is a difference in work schedules, crushed ore must be stored on surface in sufficient quantities to keep the mill going around the clock. The customary place to store the ore is in a fine ore bin next to the mill. These bins generally contain enough ore to keep the mill operating for at least 48 hours.

GRINDING CIRCUIT

Ore from the storage bins is fed, together with water, to the first circuit in the mill building. This is known as the grinding circuit and consists of one or more ball mills or rod mills. As each mill revolves, the ore within rolls over itself, causing it to be crushed and ground. The steel balls (or steel rods) in the mill assist this process.

A particularly economic form of grinding, used today wherever possible, is autogenous grinding. This process involves the use of rock-against-rock contact to crush and grind ore to the required size, thus, the cost of periodically replacing steel balls is eliminated.

Semi-autogenous grinding (SAG) refers to the addition of steel balls to the grinding stage to supplement the rock-on-rock breakage.

It is natural to expect that there will be considerable variation in the size of particles discharged from a grinding mill. Some will be too coarse or too fine for the chemical separation of their constituents to work effectively.

If the crushing and grinding process is not carefully controlled, some ore

particles get reduced to sub-micron sizes. These are called slimes and may interfere with subsequent treatment processes. That is why many crushing and grinding circuits are now controlled by computers.

Particles that are too coarse are separated from the balance of the material in a classifier and then returned to the grinding mills. While traditional classifiers consist of a box set on a slope and a mechanism for moving material up the incline, newer plants may use a hydrocyclone, which separates various sizes of particles by spinning, as in a centrifuge.

When a ball mill and classifier work together as a unit, the process is known as a closed circuit. The ground ore produced will have a certain maximum size and the amount of "fines" is limited. Usually a ball

Photo by The Northern Miner

Ball mills are used to grind ore before it is processed further.

mill grinds in closed circuit for maximum control and efficiency.

A rod mill may be used to prepare feed for a ball mill, in which case it may operate without a classifier, or, as it is termed, in open circuit.

The goal of the mill circuit is to crush and grind the ore down to what is termed its size of liberation – the maximum size to which the rock must be ground to ensure that individual mineral grains are separated from one another.

Gold Ore Processing

Throughout the centuries, gold has been recovered from its ores in many ways. These range from the rocker or long tom of the California Forty-Niner and the noisy stamp mill of the 19th century to modern methods of leaching with cyanide.

Any method of treating gold ores must take advantage of the natural characteristics of the metal. Cyanide solution, unlike most other liquids, is able to dissolve gold, and thus, is used in the processing of gold ore. When in solution (and in the presence of oxygen), cyanide slowly attacks fine particles of gold and ultimately dissolves them. It is strange, but fortunate (because cyanide is extremely toxic), that a weak cyanide solution attacks the gold particles faster than a strong solution.

For the cyanide to attack the gold particles, it is necessary that the gold first be liberated from the worthless gangue rock which surrounds it because cyanide will not attack or dissolve most other minerals.

Overall, the cyanide process is very efficient. A gold ore containing less than one gram of gold per tonne can, in some cases (and depending on the gold price), be profitably treated. A modern cyanide mill recovers or extracts 95% to 98% of the gold in the ore.

In a cyanide mill, lime and cyanide are added to the ore pulp in the grinding circuit. The lime has several functions: it protects the cyanide from being destroyed by naturally occurring chemicals called cyanicides and improves the settlement rate of the pulp in the thickening stage.

Cyanidation (the actual dissolution of the gold) begins in the grinding step. Cyanide and lime solutions are introduced here, where newly liberated gold particles are constantly being polished by the grinding action and the solutions are heated by the friction. Depending on the ore and fineness of grind, from 30% to 70% of the gold may be dissolved during the grinding process.

Additional time is required to place the balance of the liberated gold into solution. This is done by pumping the gold-bearing pulp to a number of mixing tanks, known as agitators. Here the pulp is aerated either mechanically or by compressed air, or by a combination of both, for a predetermined period of time. This varies anywhere from 24 to 48 hours. The 1980s saw a rapid expansion in gold production from low-grade oxide deposits around the world. That expansion could not have occurred without the development of a new, low-cost method of recovering the gold. That process is called heap leaching.

Heap leaching avoids most of

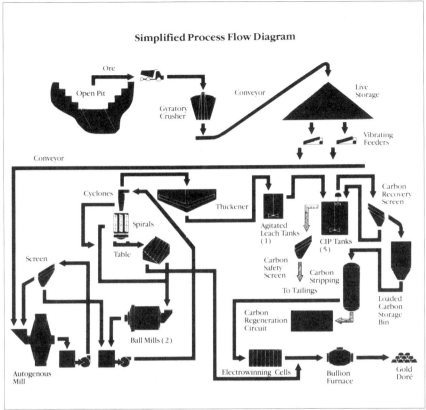

Simplified Process Flow Diagram

Flowsheet courtesy of Barrick Gold

The carbon-in-pulp (CIP) process, shown here, allows for inexpensive gold recovery.

the above steps, and does not even require that a mill be built, making it a very low-cost method of processing ore. Here, broken ore is heaped onto a thick polyethylene sheet, called a liner, and then dilute cyanide solution is sprinkled on top of the heap. As the solution trickles down through the ore, the gold is dissolved. Before the heap is constructed, the polyethylene liner is laid down in such a way that the cyanide solution will drain to a central point. From here the gold-laden solution is channeled into a man-made pond.

One downside of heap leaching is lower recovery – just 65% to 85% of the gold in the ore ends up in the gold bars a heap-leach mine produces.

TAKING GOLD OUT OF SOLUTION

Traditionally, recovering the gold from the cyanide solution was achieved by separating the gold-laden, or pregnant, solution from the barren solids present and then precipitating the gold.

The traditional approach is called the Merrill-Crowe method. The first step is to move the pulp from the agitators to one or more thickeners – large, shallow tanks. The solution

Molten gold is poured into bar molds.

Photo Courtesy of World Gold Council

flows over the top of the tank and is collected in a launder around the tank's perimeter, while the worthless rock particles sink to the bottom and are slowly raked to the centre by mechanical arms which operate continuously. This material is discharged through a pipe at the bottom of the tank but it contains too much valuable material to be discarded, so it is filtered to recover additional gold.

FILTERING

A filter is simply a large drum slowly rotating on a horizontal shaft. The drum is porous and partially submerged in a semi-circular steel tank, into which pulp from the bottom

of the thickener is pumped. As the drum rotates, a vacuum is applied, causing the pulp to adhere to the drum. Further vacuuming then sucks out the solution.

Water is sprayed on to the outside top of the rotating drum to wash out any entrapped solution. The vacuum also catches this solution. Some mill operators filter the pulp twice to be sure all that's left of the valuable gold-cyanide solution is recovered.

The remaining solid material, or filtercake, is mixed with water and pumped outdoors to a tailings pond. In the past, mill tailings were pumped into swamps and small lakes. Today, they must be adequately contained so that they cannot drain into surrounding waterways where they can damage the surrounding ecosystem. Dams and other barriers are often constructed to contain tailings.

All the gold in the ore is now contained in solutions, either from the thickener overflow or the filtering circuit. These solutions are collected in a tank and then pumped through canvas sheets to remove any fine clay particles in a process known as clarification. Clarified solutions are sparkling clear, with a light green tint. Fine zinc dust is added to the solution, and it combines with the gold to form a precipitate which is caught between leaves of canvas in a filter press.

This gold precipitate, which resembles black mud, is quite impure. It must be refined to remove the zinc and any iron, copper or other contaminants it may contain.

The modern approach is to avoid much of the above process of thickening and filtering in favor of direct gold recovery using activated carbon granules. This is called the carbon-in-pulp (CIP) process, and it is used in most newer mills because it avoids many of the solid/liquid separation stages, thereby keeping recovery costs low.

In the CIP process, the cyanide pulp is treated in four to six smaller tanks into which are added coarse, activated carbon granules (usually ground and burnt coconut shells). The gold in the solution is absorbed on to these granules and the granules containing the gold

are screened from the pulp, thereby recovering the gold.

The gold is recovered from the carbon by washing with a small amount of hot, strong sodium hydroxide and sodium cyanide solution. Gold is recovered from this concentrated solution by electrolysis, which causes it to be deposited onto steel wool cathodes. Just as in the Merrill-Crowe process, a final refining step is necessary before pure gold is produced.

Refining Gold

Refining is the most spectacular part of the process. Silica, borax and soda ash are added to the dried precipitate (or steel wool, in the case of CIP), which is heated in a furnace. This results in a miniature smelting operation. On the top of the melt is the slag containing the impurities, while the molten gold's greater density causes it to sink to the bottom.

When the furnace's contents are completely melted, the furnace is tilted and the molten material is poured into a conical mold. Worthless black slag forms on the top and is broken from the underlying gold button once it cools.

The button or buttons (there may be enough precipitate to necessitate more than one melt) are again placed in the furnace, melted and poured into bar molds.

Finally the bars are weighed, small samples are removed to determine purity (expressed as fineness in parts per thousand) and the bars, called doré bars, packed for shipping. In due course, the mine receives a cheque for the gold.

Treating Base Metal Ores

Base metal ores go through a more complex treatment process than gold to reach their ultimate commercial form. Because most base metal ores contain metallic sulphide minerals, the greatest challenge is separating the sulphur from the metal and then ensuring the sulphur is contained in some manner so that it can't damage the surrounding environment.

Unlike gold, base metals are not usually produced in a near-pure form at the mine site. This is because most mines aren't sufficiently large nor well situated to warrant the construction at the mine site of all of the plants (mills, smelters, refineries, etc.) needed to convert the metal ore into pure metal.

Instead, each base metal mine attempts to remove locally as much of the waste rock as possible from its ore and ship the enriched product, or concentrate, to a strategically situated smelter.

Crushing and grinding in a concentrator are practised in the same manner as in a cyanide mill, and for the same purpose – i.e., to liberate the valuable minerals from the surrounding worthless rock.

Flotation

Grinding of base metal ores is done in water to which certain oils and synthetic chemicals are added. Then the resulting pulp is swirled around in rectangular tanks arranged in series. These tanks are known as flotation cells. Controlled air and further chemicals, called flotation reagents, are added. The air forms bubbles in the pulp, the flotation reagents coat the metal sulphides (but not the waste particles) and cause them to stick to the bubbles which, in turn, carry the sulphides to the top of the tanks.

The sulphide-carrying bubbles are scraped from the top of the flotation machines, while the worthless material left behind sinks to the bottom and is discharged.

A recent innovation, designed to improve flotation recoveries, has been to use column flotation cells – 10-metre-tall (33 feet) tanks in which a tall column of froth can develop. Water sprayers at the top wash unwanted minerals from the froth, so that only minerals that are chemically attracted to the bubbles are retained.

Dewatering

Later, the metal-bearing bubbles, now known as a concentrate, have the water removed from them before shipment. This is typically done by pressure filters. This dewatering step is carried out to reduce the weight of the concentrate if the shipping distance is great and to prevent freezing (in rail cars, for example) during winter shipment.

By employing various flotation reagents, different kinds of metal sulphides can be floated or separated one at a time. Thus, if an ore contains copper, zinc and iron sulphides, it is possible to make separate concentrates for each metal.

Flotation is occasionally applied to gold ores, to make a gold flotation concentrate that is then treated with cyanide, with or without

Photo by The Northern Miner

Sulphide minerals adhere to certain organic chemicals. This allows the sulphides to be separated in a flotation cell while gangue minerals sink.

roasting. This method is especially applicable when the metal is very fine and intimately associated with minerals containing sulphur or arsenic. In this manner, it is possible to make, say, 10 tonnes (11 tons) of gold concentrate from 300 tonnes (330 tons) of mill feed. This concentrate can be economically treated much more intensely (finer grinding and longer agitation) than the original ore.

HEAVY-MEDIA SEPARATION

Another method of concentrating ores and minerals by floating (but actually quite different from the flotation method described above) is known as the heavy-media separation or sink-float process. Here, the differences in specific gravity of the various minerals in the rock mined are used to achieve the desired separation of ore from gangue.

In the sink-float plant, coarsely

crushed ore is fed into a suspension of finely ground ferrosilicon (or magnetite or another heavy mineral) in water. This suspension is carefully maintained at a specific gravity between that of the desired feed and the unwanted minerals or gangue. As a result, the heavier mineral components sink, and are retained, while the lighter constituents float, and are discarded.

Another way of using the density of minerals to help concentrate them is the gravity plant. In most gravity processes, a slurry of ore in liquid is fed into a centrifuge or other chamber that agitates the slurry, causing lighter particles to spin away and heavier ones to sink. Similarly, gold is often recovered on a vibrating table, where lighter materials are washed away and heavier grains remain.

Simple mechanical separation of this kind is often the cheapest method of recovery, and consequently it is most often used at the start of the process. Tailings from the mechanical separators are then sent to other processing circuits to recover more of the valuable minerals.

Heavy-media separation is also a basic step in the process of recovering diamonds. Diamonds and other heavy minerals are separated from the crushed ore, then other recovery processes separate the diamonds from the other heavies. Because diamonds tend to adhere to organic compounds, "grease tables" were once widely used to recover diamonds. Diamonds fluoresce when hit by X-rays, so separation plants are now designed to produce diamond concentrates using this property.

MAGNETIC SEPARATION

Some low-grade iron ore can be treated using magnetic separation because all iron minerals are magnetic to some degree. In this process, the crude ore, which may grade less than 30% iron, is mixed with water and ground to a suitable fineness. The pulp is then passed over a revolving magnetic drum, to which the magnetic iron minerals adhere. They are scraped off and retained, while the gangue particles are discarded with the water.

The magnetic portion is dewatered and filtered, but the concentrate produced is too fine to be used in a blast furnace at a steel plant (the destination of most iron), so it is pelletized – mixed with a suitable bonding agent and rolled on a pan or in a drum until marble-sized, iron-rich balls are formed. These pellets are dried and baked, and then shipped out to steelmakers.

Other methods of metal recovery include photometric sorting – where the distinct light-reflection properties of metallic minerals are used to separate them from gangue minerals – and various simple gravity-based processes.

Heavy-mineral concentration and magnetic separation are two approaches that have been found very useful in recovering other metals, such as vanadium and titanium, whose principal ore minerals are both heavy and magnetic. Magnetic separation can also be useful to extract low-value magnetic minerals from heavy-mineral concentrates, leaving more valuable minerals like zircon or rutile (titanium oxide) behind for recovery and sale.

Refined metal is usually produced from smelted metal in electrolytic tanks.   Photo courtesy of Vale Inco

HYDROMETALLURGY

As with other branches of the mining industry, there is an ever-constant struggle to mill ores and recover minerals and metals more efficiently and cheaply. Leaching, or hydrometallurgical, processes have been developed for the recovery of uranium and other metals.

Sherritt Gordon Ltd. developed a leaching process for producing nickel, copper and cobalt from concentrates for its Lynn Lake mine in northern Manitoba. The process employs an ammonia pressure leach, sometimes known as the Forward Process after its inventor, F.A. Forward.

In Sherritt's case, the nickel and copper were concentrated separately at the mine (these concentrates graded about 12% nickel and 29% copper). Then, the metal is leached from the concentrates by an ammonia – air mixture. Both the nickel and the copper, together with a small amount of cobalt, are dissolved and subsequently separated and recovered. The sulphur is converted to ammonium sulphate, which is recovered as a byproduct and used in the fertilizer industry.

A somewhat different adaptation of the leaching process, employing a sulphuric acid leach, is used in some uranium mills.

These processes are coming into wider use both for base metal and uranium ores. While innovative, they are not unlike the cyanidation process described earlier, but these later leaching plants are more complex and costly to construct.

In recent years, solvent extraction

and electrowinning (SX-EW) has become a popular hydrometallurgical method, particularly at copper mines. The key feature of this method is the use of special synthetic organic liquids that are able to extract copper so that it can be deposited by electrolysis. Used in combination with heap or dump leaching, this relatively inexpensive method can make it possible to economically process very low grade ores. Air emissions and waste products are generally minimal, making it doubly attractive to today's miners.

High pressure acid leaching is coming to the fore as a method for processing lateritic nickel ores. The ores are leached with sulphuric acid, then the nickel is recovered by solvent extraction and electrolysis similar to the SX-EW method for copper recovery.

Both SX-EW for copper and acid leaching for nickel enable the producer to bypass the expensive smelting and refining stages in processing the ores.

BACTERIAL LEACHING

As early as the days of the Roman Empire, the action of water-entrained oxygen on metallic sulphides has been recognized and used, particularly on copper-iron minerals. As it turns out, a microscopic organism known as thiobacillus ferroxidans is responsible for this action. This one-celled bacterium breaks down the sulphide minerals and generates weak sulphuric and sulphurous acids, which take the metal into solution as sulphates and sulphites, often leaving a precipitate of iron.

This bacterial leaching process is now used deliberately, mainly to recover copper and uranium from mine drainage waters and from surface waters percolating through old mine dumps and waste heaps.

The phenomenon of bacterial leaching has given rise to an active field of research and is producing economic benefits in the heap leaching of low-grade copper ores which are cheaply mined in open pits, particularly in the southwestern U.S.

In the low-grade, underground uranium mines of Elliot Lake, Ont., blasted ore was left in the stopes and leached with bacteria-rich solutions in situ. This reduced production costs considerably since only the uranium-bearing solution was pumped to surface, instead of tonnes of ore.

Another area where bacterial leaching is used commercially is in the treatment of refractory gold ores, in which the gold occurs as sub-microscopic particles inside the sulphide minerals. Bacterial oxidation destroys the sulphide grains, freeing the gold for subsequent cyanidation.

SMELTING

Base metal concentrates from the flotation or other physical beneficiation processes are shipped to a plant known as a smelter for the actual recovery of the contained metals. The term smelting refers to pyrometallurgical processes – those which use heat to achieve the desired separation.

Perhaps the least complicated non-ferrous smelter is one that treats copper sulphide ores for the production of blister copper, which will most likely be further refined electrolytically before entering manufacture.

Let's follow a typical copper smelting and refining process, beginning with the arrival of the copper concentrates. We'll assume the grade of the concentrate is about 26% copper, linked with iron and sulphur in the form of the mineral chalcopyrite. The first step in the smelting and refining process is to roast the concentrate.

Roasting involves heating the concentrate to a very high temperature while in contact with air, or oxygen-enriched air. This burns off part of the sulphur; more importantly, it changes the copper-iron-sulphur complexes to chemical forms which are more amenable to the smelting or reduction process. In roasting, some of the sulphur actually acts as a fuel and, in some roasters, little or no additional fuel need be added once the system has been brought to incandescence (when the heat causes it to glow). The product of the roaster is known as a calcine, and in our example its copper grade would be increased to about 31%.

The calcine is mixed with various reagents known as fluxes. Different fluxes are used for ores of different mineral composition. For instance, silica is used for an ore high in lime.

The flux-ore mixture is fed into a reverberatory furnace – a long, flat chamber, in which a flame is shot from one end to the other, where a flue system removes the hot gases. In the furnace, the fluxes react with the gangue minerals to form various low-melting silicate minerals known here as slags.

The copper, and much of the remaining iron and sulphur, form a matte. The matte also picks up and dissolves any precious metals that may be present. In the furnace, which operates at temperatures above 1,100°C (2,000°F), the slag floats on top of the heavier matte and is tapped off periodically to be sent to the dump.

The molten matte, now grading about 46% copper, is also tapped off and poured into a converter, together with more fluxes and reducing agents, where it is blown with air. In this final furnace treatment, the iron becomes fully oxidized and unites with the fluxes to form a slag. The copper becomes reduced to its elemental state, and the last of the sulphur is driven off in the form of

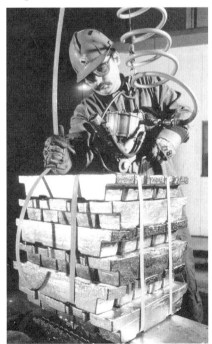

Photo courtesy of Teck Resources

Zinc ingots are shipped to market for use, primarily, in steelmaking.

sulphur dioxide. The iron-rich slag floats on top of the molten mass and is periodically sent back to the reverberatory furnace until it no longer contains any copper.

The converter produces blister copper, which is about 99% pure.

REFINING

After a further, slight fire-refining, the copper is cast into shapes known as anodes. These are shipped to an electrolytic refinery for the purification of the copper to commercial specifications, and for the recovery of the precious metals the copper has gathered to itself in the converter.

In the electrolytic plant, the copper anodes are placed in tanks containing copper sulphate. Thin sheets of pure copper are also placed in the tanks to act as cathodes. Electric current is passed through the system of anode-electrolyte-cathode, and the copper is carried from the anodes to build up on the cathodes in a highly purified form.

Eventually, the cathodes are removed and melted down for casting into various commercial shapes. Any precious metals in the copper anodes fall to the bottom of the tank with the last of the impurities and become contained in a muddy deposit, whence they are recovered and refined in a separate process.

The smelting of complex, nickel-copper sulphide ores and the refining of their contained metals involves a much more complicated suite of processes. Nevertheless, the same, general ideas outlined above for the smelting of copper are to be followed.

In the case of lead derived from galena (lead sulphide) ores, a lead concentrate is roasted in a sintering machine, then fed to a blast furnace along with coke, which acts as a fuel and as a reducing agent, and various fluxes. The actual reduction to metallic lead is carried out entirely in this furnace, which produces an impure lead bullion. The bullion is refined electrolytically and the precious metals contained are recovered in much the same way as in refining copper.

The treatment of zinc concentrates follows the pattern of roasting to a form in which all of the sulphur has been driven off, and the resulting calcine consists of zinc oxides and sulphates, iron oxides and sulphates and various gangue minerals.

The calcine is leached with sulphuric acid to dissolve the zinc as a sulphate. The leach solution is purified chemically so that a pure zinc sulphate solution may be fed to an electrolytic tank for its final reduction to the pure metal.

8. Mining and the Environment

SAFEGUARDING THE ENVIRONMENT
Mining companies supply the metals and minerals that humanity uses for shelter, survival, work and pleasure, as well as the expansion into space and interplanetary endeavors. At the same time, they want to conduct this business in an environmentally responsible manner. Yet mining by its very nature requires that land, air and water systems be disturbed. While the economic benefits of the industry are as important today as they ever were, the public has become increasingly concerned about the impact that mining is having on the natural environment.

The metals and industrial minerals that mining produces can find their way into the environment and become pollutants. The byproducts that occur with the metals, such as sulphur and arsenic, can be dangerous to the environment if they are released. The fuels and chemicals the industry uses to do its job are also potential pollutants. Mining creates and employs hazardous substances that must be handled with care.

Other pollutants produced by the mining industry are of concern more to the workers in the industry than to the public at large. Dusts, for example, which are often hygienically hazardous, are produced by many mining activities. Noise, too, is a form of pollution of concern for those in the work environment. In uranium mines, the products of radioactive decay are a major concern.

The challenge for companies is to find, extract and process mineral resources with the least possible disruption to the environment. To meet this challenge, they adopt a broad range of protective measures, including: sensitive treatment of the land during exploration; environmental and aesthetic management of land under development; environmentally sustainable pro-

Steps in Environmental Planning for Mines

Exploration stage:
- baseline environmental studies
- initial public notice

Feasibility stage:
- select mining methods
- obtain initial approvals

Development stage:
- design tailings ponds
- design wastewater systems
- make closure plans
- inal approvals

Production stage:
- give notice of operations
- monitor air, water and soil quality
- waste management
- maintain wastewater systems
- monitor tailings ponds and waste piles
- post bonds for closure

Mine closure:
- cap tailings ponds and wastepiles
- demolish and remove structures
- complete cleanup
- continue monitoring water quality

duction procedures during the mining and metallurgical processes; and decommissioning and reclamation practices aimed at restoring the land.

Environmental performance and accountability are important issues for mining companies, their shareholders and the public. Most companies now include a discussion of environmental issues in their annual reports so as to keep shareholders and the public informed about the steps they are taking to protect the land, water and air quality at their operations.

THE BAD OLD DAYS

Society hasn't always kept its house clean. Old mining operations frequently dumped wastes without concern for their physical or chemical stability or used milling and smelting techniques that released pollutants into the atmosphere, lakes and rivers. Some gold ores, for example, used to be roasted. These were heated until sulphur and arsenic in the ores were driven off as gases, which were released directly into the atmosphere. Another technique, the amalgamation process, used mercury to extract gold from ore. The mercury was then boiled off, leaving the gold behind. This process released mercury – one of the most environmentally hazardous of all the metals – into the atmosphere and often let it enter soil or water through spills. The poorly controlled burning of fossil fuels to run mills and fire smelters also took its toll on the environment.

As people have begun to realize that fouled environments are unproductive and hostile ones, environmental controls have become stricter, and governments have taken a role in making sure that industries don't make the mess they once did. Most industrialized countries have regulations governing air emissions, effluent discharges to watercourses and the disposal of solid wastes. Many less-developed countries, although they sometimes feel themselves faced with a choice between economic growth and a safe environment, are adopting stricter regulations too.

Both stricter regulations and the knowledge that environmental responsibility serves everyone's interest have prompted mining companies to develop their own codes of practice to ensure that mining operations do not significantly harm their surroundings. Their goal is to adhere to these standards both at home and abroad.

BEFORE MINING BEGINS

Environmental protection begins at the earliest stages of mine exploration, long before the first ore is extracted. During this stage, companies make an effort to minimize the impact of prospecting, drilling, trenching, road building and other related activities. Exploration activities usually affect the environment only temporarily and, with proper planning, work can be carried out with minimal disturbance to land, vegetation and wildlife habitats. Even so, companies have learned that it is important to keep local communities informed about their activities. This consultation process sets the stage for good community relations once mine planning begins.

To keep public support, mining companies must demonstrate respect for the ecosystem in which they are working and adopt a broad range of protective measures. The drilling fluids and lubricants used in diamond drilling can seep into the water used to bring cuttings to surface. This water must be properly contained and disposed of so that it does not contaminate the groundwater. Drill holes often have to be sealed with impermeable concrete or bentonite (a clay material) to ensure that the drill hole cannot act as a channelway for contaminants to reach the groundwater from surface.

Another consideration in mineral exploration is the safe handling of camp wastes. This means more than just being careful not to litter, as isolated exploration camps must ensure that they handle fuels and dispose of human wastes in ways that do not contaminate the natural environment.

In very sensitive areas, such as tundra regions, it is common for governments to require that exploration crews have permits to work, setting down limits on what the crew can do. Respect for wildlife must be shown at all times. With proper planning, forethought and good housekeeping, all of the impacts of an exploration campaign can be minimized. In recent years, a major exploration-industry association, the Prospectors and Developers Association of Canada, developed a set of practices called E3, for "Environmental Excellence in Exploration." A distillation of the knowledge that had been built up by mineral explorers over the years, it gives explorers a blueprint for doing their work in ways that limit the effect on the environment.

Once a deposit of economic interest has been outlined, studies and sampling programs are carried out to provide data that are used to shape a project's design. Specialists research all aspects of the environment to establish basic data, against which future test results will be compared and evaluated. A few of the many areas investigated are: soil composition; the concentrations of metals in nearby watercourses; the populations of animal and plant species that live nearby; air quality and climate; historical and cultural sites; and numerous other pieces of data that allow regulators to determine whether the mine, once in operation, is causing adverse changes to the environment.

The Permitting Process

Once a promising orebody has been found, most countries require a technical environmental study of the proposed development. These reviews are so detailed that they can take several years to complete. In order to get a permit for mine construction, companies must provide details of their operating plans, as well as the results of engineering, environmental and socio-economic studies. A further step before permits are issued is often a series of public hearings that allow individuals to voice their concerns or support. Needless to say, obtaining permission to develop a new mine can be complex and time-consuming.

The permits spell out the terms under which an environmentally acceptable mine may be developed. Typically, a government agency responsible for environmental protection or mining development will inspect the study and permit or prohibit the mining operation based on the study's findings.

One concern is the simple physical stability of the mine workings – whether the pit walls will stay up or whether mine workings will cause nearby ground to subside. The other concern is that mining may allow contaminants to enter the atmosphere, surface waters, soils or the groundwater systems. Mines, mills, smelters, tailings ponds, effluent discharges and smokestacks must be designed to keep contaminants out of the surrounding environment.

Base metal ores are usually natural compounds of a useful metal with sulphur. In the presence of oxygen and water, the sulphide minerals react to form sulphates and sulphuric acid. These compounds are not highly concentrated, but contain sufficient acid and salts of heavy metals to be harmful to aquatic life in surface watercourses. Because

Photo by The Northern Miner

Care must be taken when a foreign mining company moves into an area where local artisanal miners are active, as they may view the mining company as a major threat to their livelihoods.

coals also contain iron sulphide minerals, the same chemistry happens in coal mines.

Careful control of mine waters and surface runoff allows mines to prevent or minimize the effect of acid-generating waste rock and tailings. In difficult cases, the mine may channel its runoff water to treatment plants that neutralize the acid waters before they are released to surface watercourses.

Water also leaks into mines from the surface and through the groundwater system. This water is collected and pumped to surface, and needs to be disposed of. However, it can be slightly acidic from its contact with the sulphide minerals in the mine or can carry heavy metals that it has dissolved from the mineral deposit. To keep the mine from polluting its surroundings, this water may have to be treated to remove the contaminants before it is discharged to lakes or rivers.

There are several ways to treat such waste waters. Lime or other alkaline compounds can be added to neutralize sulphuric acid, and the heavy metals can be extracted chemically. Most countries have limits on the concentration of contaminants waste waters can contain if they are to be released to surface watercourses. These waters are treated so that any contaminants are present in concentrations lower than the prescribed limits.

Mines also have to dispose of waste rock. It comes up every day of a mine's life. Some can be used as backfill, but there is usually plenty left over when the mine is finally exhausted. Mines have to be designed with waste rock dumps that will not slide or collapse. The waste rock also may contain traces of mineralization, and the contact of rain or snow with metal sulphides can form acidic runoff that can poison nearby watercourses.

The best way to prevent acidic runoff from reaching watercourses is to prevent it from forming in the first place. This is done by diverting water away from the waste dumps and by capping the dumps with impermeable soils that prevent rain and snow from reaching the waste rock. If that solution isn't practical, then the runoff that forms can be collected and treated to lower concentrations of contaminants to a safe level before the water is discharged.

The permitting process and hearings can take one to three years. Applicants frequently amend their project plans to reflect local concerns or take advantage of cleaner technologies. The result is an economic development without lasting environmental disruption.

MILLING, SMELTING AND THE ENVIRONMENT

After the ore has been brought to surface, the process of getting the metal out also can create harmful substances. Apart from the ores they use and the wastes they produce, milling and smelting run on fuels and use chemicals to extract the metals. These substances also can present a hazard to the environment.

The milling process uses plenty of water, and large mills may use several hundred litres a minute. This water contains small concentrations of vari-

ous organic and inorganic reagents used in the milling process. Many companies recycle all or part of the water back into their mills instead of discharging it into the environment. If it has to be discharged, then it also must be treated by capturing or destroying the chemicals.

One particularly important example is gold extraction. Most mills extract gold using weak solutions of sodium cyanide under slightly alkaline conditions. Traces of cyanide are left in gold mill effluents. Cyanide is not a single chemical element, so it can be broken down into the elements that make it up, namely carbon and nitrogen. The waste waters from the mill can be held in a pond where sunlight and contact with air break the cyanide down. Another choice is to incorporate a cyanide destruction process into the mill circuit, in which waste waters are aerated or chemicals are added that react with the cyanide to form less hazardous substances.

The use of cyanide in mining sometimes raises public concern. It is a dangerous poison and can be fatal if ingested. But since cyanide is also a compound, and an unstable one at that, it is relatively easy to destroy any excess in waste waters, leaving them suitable for discharge to the environment.

TAILINGS

Mills also produce solid wastes, tailings, made up of the finely ground rock that has been separated from the ore minerals in the milling process. The tailings can contain hazardous byproducts the milling process has

Photo by The Northern Miner

New exploration or mining activity near established settlements can rapidly bring locals into conflict with mining companies if good relationships aren't forged early.

separated from the useful materials. For example, gold ores often contain arsenic and uranium ores almost always contain radioactive decay products like radium. Tailings are disposed of in tailings ponds that must be designed to keep these byproducts where they belong, preventing them from reaching the environment. This can mean adding a treatment process that captures the byproducts chemically so that they can then be safely disposed of separate from the main portion of the mine's tailings.

Often the tailings contain worthless sulphides, such as the iron sulphides pyrite and pyrrhotite, that were present in the ore along with the useful ones. Like those in the waste rock, sulphides in the tailings must be kept from forming acid runoff. Again, this means diverting water away from the

tailings pond or treating the water from which it drains.

Most iron ores have little or no sulphur. In iron mining, the main concern is not acid mine drainage, but the water discharged from the gravity and magnetic concentration processes and from pelletizing plants. This water often contains particles of stable iron oxides. These minerals have a high specific gravity and tend to settle out of water rapidly, affecting waterways only short distances from the point of discharge.

When the tailings ponds are finally full, they must be capped with impermeable soils such as clay to prevent water from percolating through the tailings and drawing potentially harmful substances into the groundwater. The capping material is usually seeded so that plants will grow on the surface and prevent wind and water from eroding the cap.

SMELTER GASES AND DUSTS

When sulphide ores are oxidized during the smelting stage, huge amounts of sulphur dioxide are produced. If the sulphur dioxide goes up the smokestack of a smelter, it can react with the water vapor in the air to form sulphuric acid, the acid in acid rain or snow. It is estimated that about 60% of all sulphur emitted to the atmosphere comes from smelting and other industrial activity. Ores that can be treated by hydrometallurgical methods do not produce air emissions, but process waters may have to be treated before they can be discharged.

Many smelters now trap their sulphur dioxide to make sulphuric acid, which is a useful product in itself. A large smelter can produce thousands of tonnes of sulphuric acid a day. The gases are drawn through a catalytic oxidation process that turns the sulphur dioxide into sulphur trioxide. Dissolved in water, this chemical reacts to form sulphuric acid which can be sold to the general bulk chemical trade or used to make agricultural fertilizers.

Dusts in the smelter exhausts can be recovered by passing them through cyclones and electrostatic precipitators. These units can recover up to 95% of particulate dusts, returning them for processing to recover their metal content.

AFTER MINING ENDS

Mining is always but a temporary use of land, and an important goal of the operating company is to return the mine site to a natural and stable state, thereby making it available for other uses.

Towards this end, the industry has embraced mine reclamation techniques that consist of removing, relocating or demolishing buildings and physical infrastructure; closing pits and shafts; stabilizing underground workings, soil and slopes; treating tailings and waste water appropriately; and revegetating land.

Reclamation can be carried out at various stages of mining activity – after exploration; after surface or underground mining; or after treatment and processing facilities have been closed.

9. The Mining Team

MINE FINDERS

As a result of advances in technology, mineral exploration has changed dramatically from the days when the lone prospector packed a pick and a rabbit's foot into a canoe and headed into the bush for a season's work. Mineral exploration and mining is now a business that calls for highly skilled individuals to work as a team, using powerful, often computerized, exploratory and mining equipment. The exploration team can include prospectors, geologists, geophysicists and geochemists (and their assistants), whose skills complement each other as they search for new mines.

Prospectors still play an important role in generating showings (evidence of local mineralization), which are later optioned and explored by mining companies. To find these showings, prospectors rely on geological maps, government reports, assessment files and aerial photographs.

Government geologists lay the groundwork for future discoveries by conducting regional-scale programs and by preparing reports and maps from the data gathered. The release of new information by government geological surveys is, therefore, eagerly anticipated by keen prospectors. Exploration is a competitive business, so having a jump on the competition can make the difference between making a discovery and missing out on one.

Prospectors test promising areas by early-stage field work, which might include following a train of mineralized boulders to their source or collecting samples from soils and rocks to identify and test anomalies. Old-fashioned "boot-and-hammer" prospecting is still an important tool in mineral exploration and has led to many spectacular discoveries, including the Voisey's Bay nickel-copper-cobalt deposit in Labrador. Because prospectors play such an important role in finding new showings, some governments offer small grants to encourage their continued efforts.

Prospectors often vend their

Photo courtesy of Royal Oak Mines

After a long day's work in an underground mine, mine workers head for home.

properties to mining companies, which will then send out a team of geologists to carry out more detailed sampling programs. A geophysicist searches for variations in the physical characteristics of the earth that may be caused by the presence of minerals. A geochemist analyzes the metal content in rocks, soils, surface waters or plants, searching for anomalous values that differ from background metal levels in the region. Usually, more than one technique is applied to check any anomalies that are identified. Trenching or pits may provide some early samples of mineralized rock for testing.

Advanced properties see the arrival of diamond drillers. These men and women spend much of their time in field camps and are accustomed to moving from job to job since diamond drill contracts seldom last long.

In all but the smallest field camps, one of the most important members is the camp cook, whose offerings play a large part in sustaining the morale of the crew.

THE MINERS

Once the mine is in production, all of the various mining functions require specialized equipment operators. All operators fit the bill as miners, but individual roles include drillers, muckers (mucking machine operators), blasters (who use explosives to break rock), rock-bolters (who insert rock bolts to support a mine's ceiling and walls), pipefitters (who string the pipes that supply water and compressed air to the drills), electricians, carpenters, maintenance crews and so on.

Underground mining skills are not unlike those required in the more familiar industrial trades or technical professions.

When the mine is on surface, only the equipment changes (and it's generally much, much bigger). The teamwork is the same. Open-pit miners operate huge bulldozers, loaders and haulage trucks carrying several hundred tonnes in each load, giant shovels which scoop up tens of tonnes in a single bite and massive rotary drills for drilling the blast holes needed for open pit mining.

Work Wear

Before the beginning of his or her shift, which usually lasts 10 hours, the miner reports to the dry or change house. Street clothes are exchanged for work clothes and then there is time to chat with friends while waiting for the cage to go underground.

Warm underclothes, a pair of overalls that are lightweight but strong, steel-toed rubber boots, and gloves are standard work wear. The miner straps a heavy safety belt around the waist to support a battery pack for a head lamp. A hard hat is worn, to which the lamp is attached.

The miner also wears safety glasses, and ear plugs or ear muffs for protection from excessive noise. If the mine is wet, the miner will wear a set of waterproof outer clothing often referred to as oilers.

Health and Safety Are Key

The men and women employed in mining must be in good health. Although mechanization has reduced the amount of physical work required, the individual must still have reasonable strength and good hearing and eyesight. A physical examination is required by law upon starting employment in the mining industry and medical check-ups are carried out on a regular basis.

The development of safe job procedures, combined with common sense, has always been a high priority for mining companies. Accident prevention is the responsibility of everyone involved – management,

Photo courtesy of Camborne School of Mines

Mine workers are given both classroom and on-site training in mines.

workers, unions and governments.

Each mine also has a rescue team. These usually consist of six to eight miners specially trained to find and assist trapped miners in the event of cave-in, fire or some other accident.

Powerful modern ventilation systems have greatly reduced the risks once associated with mining, but mining companies, unions and governments continue to work with research organizations in order to ensure that working conditions are always improving.

Skill Level Increasing

While some unskilled workers with little post-secondary school education still work in North America's mines, the increasing complexity of the industry requires that those who wish to advance to the highly skilled and better-paying jobs should have some post-secondary education. Technical training from a technical high school or community college is an advantage and higher education such as an engineering or other university degree gives a young miner many more career possibilities. Mining companies provide the inexperienced worker with initial training common to all employees. This common core training (or stope school, as it is sometimes called) includes an introduction to the basics of mining and mining safety procedures, classroom study on surface and underground, followed by on-the-job training as helper to an experienced miner. Specialized training is then necessary to become qualified to operate individual machines.

Office Staff

Besides the miners and various supervisory personnel, there are the technical and professional staff whose duties encompass such things as sampling, surveying, drafting and planning. Directing their activities are the mining engineers and geologists who map the progress of the mining operations, design the mining methods, and direct the search for new ore. Upper management includes a mine manager and a mine superintendent, plus accounting and executive-assistant personnel at each mine.

Mill and Smelter Workers

The treatment of ore demands another set of skills, and thus, another group of specialized operators are on the job to watch over the ball mills, flotation tanks and other machinery. Assayers and chemists are also on hand to carry out the analysis of samples for the control of the milling and mining operations.

When the metal concentrates are shipped out for refining, yet another team of workers comes on the scene – furnacemen, smelting equipment operators and refinery operators, to name a few.

Computerized "programmable logic controllers" are common in most mills and smelters, so operators are required to monitor a whole sequence of operations from a control console that keeps tabs on what is happening in various parts of the plant.

Finally, the gold, copper or other metal is ready for shipping, making the shippers the last workers in the unbroken line of people who discovered, mined and processed it.

10. The Business of Mining

Claim Staking

There can be no healthy mining industry without a secure and fair system of land tenure.

A system where people determine the ownership of a carload of ore by trading gunfire until one or the other gives up is effective enough, but dangerous to the public peace. Much better are systems under which the law determines the ownership of mineral rights.

The first concern of any successful minefinder is to be assured of the benefits from his or her efforts. Mineral laws enable the prospector to do just that.

Although these laws vary from one country to the next, many of the same principles apply. In most countries, the intent and spirit of the law is:

- to secure for persons and companies the exclusive right to pursue development of a mineral discovery;
- to protect the public interest;
- to encourage exploration and prevent owners from tying up ground without exploring it – to use it or lose it;
- to provide the means whereby disputes may be settled quickly and at minimum expense.

There are two widespread systems. Claim staking is a system that allows a prospector – whether an individual or a company – to establish a right to mine in a certain area. It is the usual way of establishing mining rights in countries with legal systems deriving from English common law. In some places it is done by placing physical marks or monuments on the land itself, and then reporting the act of staking to the government.

More and more jurisdictions allow "map staking," where the prospector simply applies for the right to mine an area without physically staking the ground. In either case, the prospector gets the right to the mineral resources of the land only if he or she is the first to apply for them.

Permit System

The other widespread system of land tenure is the permit system. In this scheme, the government controls the mineral rights and licences the prospector to explore a certain area. The permit – also called the concession, licence, or contract area, expires after a specified period; usually the prospector can renew the permit but must drop part of the area it covers. This provision ensures that the holder works continuously on exploration in order to know which parts of the contract area to keep at the next renewal. The exploration permit may also specify minimum amounts of work that must be done, or money that must be spent, on the area to keep the mineral rights.

The claim system, because it allows prospectors to stake open ground without requiring applications or prior agreements, rewards companies and individuals that move quickly to pick up mining rights. The permit system's requirement for a formal exploration agreement rewards large groups with the backing to carry out the plans.

Both the claim system and the permit system give the holder the exclusive right to explore and develop an area. To keep that right, the holder is required to perform work; if the work is not done, the ground falls open for someone else.

It is also usual that holders of mining property must submit technical reports to the government as proof that the exploration work has been done. The reports are opened to the public and become useful information for future prospectors.

Photo courtesy of Aur Resources

In parts of Canada and several other countries, mining claims are marked with claim posts.

Raising Capital

You don't have to be rich or work for a large mining company to find a mine. But there is one essential ingredient for any successful mine exploration and development program – access to enough funding to see the program through to completion.

Mining is a business like any other – with one exception: the assets of a mine are non-renewable. That means that during the course of production, the primary asset that gives a mine its being is consumed. Because the assets are non-renewable and disappear with time, mining is a risky business. The risks are greatest at the beginning of the life of the enterprise, when geological information is sparse. Consequently, for the investor, this is the time when the greatest gains (and losses) can be made.

To see how it works, we'll follow the financial history of a hypothetical exploration/development company called Morty Mines Ltd. Beginning with the pre-discovery phase, we'll follow it to the point where it matures to become a successful commercial mining operation. As Morty Mines evolves, it will look for financing in different places.

Pre-discovery

Every mineral discovery requires not only money, but a lot of dedicated effort from many different sources. Many discoveries result from the efforts of prospectors who are financed by a grubstake. This is an informal business association between friends or associates who

put up seed money to do some initial work with the intention of participating in the benefits. That is to say, a friend may finance the prospector's work in exchange for an interest in whatever he or she may find.

Discovery

Let's assume that our prospector, Morty, has staked some ground and found a surface mineral showing. Most prospectors or grubstakers can only put together enough money for a limited amount of work on their ground – some sampling or trenching, maybe a few days work with a backhoe, or one of the less expensive geophysical ground surveys like magnetics or VLF-EM. Beyond this point, they have the choice of forming a company to do more work or finding an established company to take on the work – and the expense.

Under an option agreement, Morty offers an interest in the property to a company with enough funds to do further work. In exchange, the company commits to spend a specified amount on exploration – when it has done that, it has earned its interest. Often the company taking the option will make a cash payment or issue some of its shares to the property-holder.

The original holders may be left with a participating or working interest in the property, in which they fund part of the future work, or with a carried interest, which gives them a royalty on future production without the obligation to contribute to future expenses.

The other choice for Morty and his grubstakers is to form a limited liability company. Their company could then issue a definite number of shares to raise the capital it needs.

In the case of the incorporated company, consideration for the money and effort our grubstakers have put into the discovery will be returned to them in an agreed-upon number of shares in the new company – Morty Mines Corp. These shares are called vendor shares and the vendors will not be allowed to turn around and sell them in the market. Rather, they will be held in escrow, usually in the hands of a trust company acting under instruction, until a successful application may be made to the securities commission for their release, to be traded on the stock market. The reason for this will be apparent shortly.

Equity Financing

If, for instance, Morty Mines is capitalized at five million shares, and one million shares are pooled or placed in escrow for the vendors' interest, the treasury will contain four million shares the company can sell to raise funds. A financial or brokerage firm is commissioned to sell the shares to its clients in what is called an initial underwriting.

The firm usually does this by buying the four million shares, then issuing a prospectus to its clients. This is a written document which includes all the financial and technical details of the company and its properties. The prospectus allows the company to apply for a listing on a public stock exchange, where the investors can trade their stock.

Small exploration companies generally get their first listings on exchanges that specialize in trading the shares of venture-capital companies, such as the TSX-Venture exchange in Canada, the NASDAQ in the United States, or the London Stock Exchange's Alternative Investment Market. If they become successful, they will "graduate" to the major stock markets.

Now, if all of the vendor shares, or any substantial part of them, were to appear on the market without warning or control, the underwriter would be severely handicapped in his attempts to sell the shares, for which he has put badly needed funds into the Morty treasury. This is why the vendor shares are held in escrow – to protect the underwriter.

Development and Production

Let's now suppose that our exploration program has found a major deposit. Production is justified at a rate of 1,500 tonnes per day. Morty Mines may still have a million shares in the treasury, and perhaps a small cash balance. But it is now faced with the problem of raising something like $150 million for the construction of a major mine and mill.

Morty Mines is not as risky an investment anymore; it has ore in the ground and a good chance of making money. This gives the company more ways to finance the development of the mine.

In rare cases, it may be possible to raise a substantial part of the money the company needs by selling the million shares remaining in the

Illustration by Rob Loree

Joint ventures with major companies help junior firms finance their exploration work.

Morty treasury. Market conditions and patterns of investor behavior at the time may be such that the sale of these shares (i.e. equity), together with a bond or debenture issue (i.e. debt), may fit the bill. There are other possibilities: for example, it may be possible to combine equity with debt financing.

Often, however, it will be necessary to create a large number of new shares. The simplest way to do this is to reorganize the company. In this case, a new company, New Morty Mines Corp., will be formed. It will have its own share capital structure and its own identity separate from that of its predecessor. The property and assets of Morty Mines will be transferred to New Morty Mines in exchange for some agreed portion

of the New Morty shares, to be distributed among the equity holders of the older company. These shares in turn may also be pooled or placed in escrow in the same way and for the same reasons described above.

New Morty Mines then sets out to sell its shares through an underwriter. If our example is as good as we set out to make it, New Morty will be able to raise the necessary funds to bring its find into production.

DEBT FINANCING

If conditions are right, Morty Mines may get debt financing, by issuing a bond or a debenture and selling units in it to the public. The bond or debenture is treated as a loan and must be repaid when the company starts to make money.

In some cases, the bond or debenture certificate may have a warrant or right attached to it. This is in order to give the investor an added inducement to purchase the bond.

A warrant usually gives the purchaser the right to buy a certain number of equity shares at a stated price if exercised by a stated time. The warrants or rights are themselves negotiable on the market, and often listed on a stock exchange.

Banks are often willing to consider project loans to a mining company for a specific mine, or for an expansion. But the mining company must have a sure orebody and management with a proven track record.

One way to borrow money at low interest rates is to pre-sell some of the metal from your future mine. For example, a company with a potential gold mine in the works, can borrow bullion from a financial institution in exchange for bullion from future mine production. The company sells the bullion at the prevailing market price and uses the cash to build the mine. Once in production, the company then pays back the loan with actual bullion produced from the mine. This is called a gold loan, or "hedging" your mine production.

If markets are weak or the project is very large or complicated, Morty Mines may decide to form a joint venture with a senior company to develop the mine. Typically, the major company will put up the capital required to build the mine in return for a direct ownership interest. This may range from 25% to 60% or more, depending on the project's economics and the capital cost requirements.

In certain cases, the major company may make an offer to buy all the shares of Morty Mines, sometimes at a premium to the trading price. In this way, a patient shareholder can be rewarded by receiving a return many times his original investment.

11. Feasibility: Does it Pay?

WINNERS AND LOSERS

Mining is a large, vital and often lucrative business. Its rewards are spread across a cross-section of our population. But not all mining ventures are successful. Risks are high and they take many forms.

The process of discovering and developing any mineral deposit involves dozens of different people with different skills, and the expenditure of many millions of dollars. But the question to ask when evaluating a deposit is always the same: Does it contain enough recoverable and marketable metal or gems to be dug out of the ground, transported to market and sold at a profit? Obviously, there are risks involved in each of the steps, and one wrong calculation can be disastrous.

The most serious risks in any mining project are those associated with geology (the actual size and grade of the mineable portion of the orebody), metallurgy (how much of the metal can be recovered) and

economics (metal markets, interest rates, transportation costs). But there are many others, such as problems arising from unforeseen political developments, new restrictive regulations or the availability of workers, to name a few.

FEASIBILITY STUDIES

One of the features that distinguishes a mining enterprise from many other businesses is that during production, the company's asset (i.e., the ore) is progressively consumed. Some day, the mine's assets will be gone, hence, a mine is referred to as a wasting asset. This has important implications for the justification of allocating capital to any new mining project.

The time value of money plays an important role here. Simply put, the annual profits generated by a mine must be sufficient to pay back (within a reasonable time) the money invested in the mine. It is the job of mining engineers to estimate the "payback period" in what is called a feasibility study.

One of the important elements in a feasibility study is the estimate of mine operating costs. It is impossible to suggest what the costs might be for a particular mine without looking at all the details of the planned operation, and reasonable estimates can only be made when complete information is available. The final estimate will only be as dependable as the information used to arrive at the individual cost estimates from which it is derived.

The prices the mining company will have to pay for labor, electrical power, supplies and the shipping out of its concentrate are all factors that influence the capital costs of a mining project.

Each country has its cost advantages and disadvantages. For example, mining in the vast, undeveloped regions of northern Canada makes the construction of roads, railways and airstrips much more expensive than in warmer climates. Also, miners in Canada and the United States demand higher wages

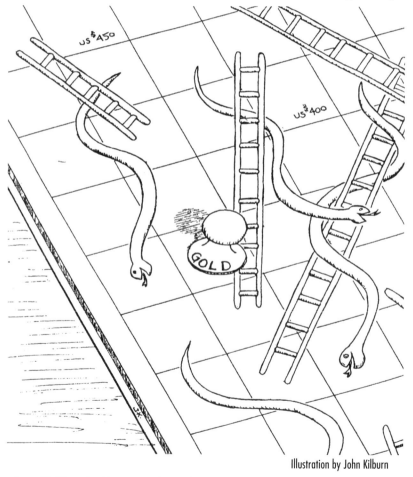

Illustration by John Kilburn

The snake and ladders of gold mining: a drastic change in gold price can suddenly make a marginal project profitable – or uneconomic.

than their counterparts in developing countries.

On the other hand, mining companies working in many developing countries can encounter problems such as high tax and tariff costs, and the corruption of civil servants such as customs officials, without whose help they would have difficulty getting their project off the ground. The overall political instability of some countries can be a great deterrent to the development of mines.

Somewhat perversely, however, the existence of any combination of negative factors leads to less exploration in that country or region, which, in turn, can increase one's chances of discovering an economic orebody. In mineral exploration, something is always better than nothing.

GEOLOGICAL RISKS

Geological conditions pose technical challenges for mine-operators as well. The huge copper-zinc orebody in Flin Flon, Man., for example, was not developed until many years after it was discovered. It was delayed because in 1915, when the deposit was discovered, there was no economic way to extract the zinc from the ore, which contains more than 15% talc.

Also, many mineral deposits are oriented at awkward angles. This can cause difficulties in handling the ore underground and in supporting the ground while it is being mined.

Another risk is the quality of the work done to calculate reserves in any given deposit. This is particularly true of underground deposits, which may pinch and swell and be less continuous than indicated by surface drilling and initial deposit modelling. To mitigate this risk, mining companies often carry out an underground exploration program, which may include closely-spaced, infill drilling to upgrade resources into reserves and, possibly, a bulk sample to get a better handle on grade. Usually the reserve calculation is audited by an independent engineer in order to reduce the risk for mine financing.

Some mines run into ground support problems when mining reaches a considerable depth. Those problems can increase the complexity and cost of mining. While in the early years of a mine's life, extraction may have been relatively inexpensive, but costs may escalate as mining progresses at depth, so this must be anticipated and accounted for in the feasibility study.

THE GROSS VALUE FALLACY

Mining companies and investors often make mistakes when calculating the value of mineral deposits. One of the most common errors is a simple calculation wherein reported percentages of metals are multiplied by the current market prices for those metals, yielding what is known as the gross value of the reserves. This figure is virtually meaningless.

Consider a lead-zinc orebody with reserves grading 8.5% zinc (85 kg of zinc per tonne of ore) and 4% lead (40 kg per tonne). If the current market price for zinc is 75¢ per lb. ($1.65 per kg) and for lead is 28¢ per lb. (62¢ per kg), a simple computation suggests that the ore has a gross value of $165.05 per tonne. But this calculation is misleading, and can, in some cases, be downright dangerous. The costs of transporting, smelting, refining and marketing all affect the payment made by a smelter to the mine. Premiums for valuable byproducts mined can be a bonus, but penalties that a smelter charges for elements that present metallurgical or environmental problems (like mercury and arsenic) can be costly.

After these deductions, we may find that our ore has a value of $65 per tonne, a drop of more than 60% from the gross value figure of $165 which had once seemed so promising. And it is this $65 received by the mining company that must cover the cost of mining, milling, administration and taxes, as well as provide a profit.

Studies have shown that a typical base metal mine will generally get about 45% to 65% of the gross metal value as a net return from the smelter. Some complex ores, which will tend to have higher smelting costs, can return as little as 35% of the gross metal value.

Gold mines differ somewhat in that smelter charges do not apply to simple gold ores (unless we are considering the gold content of a base metal concentrate). This is because a very high percentage of the gold contained in the ore can usually be recovered at the mine site without the need for shipping a concentrate to a third party. Typically, as much as 85% to 95% of the gold contained in a gold ore can be recovered at the mine site and a doré bar can be produced on the premises. The only transportation charges involve the cost of sending the gold bars to a refinery or mint, whose charges are minimal. The same applies to the ores of silver and other precious metals.

THE MINING RATE RISK

Economies of scale operate well in the mining industry: a big mine will produce significantly more output per unit of input than will a small mine. But not all orebodies can support a big mine.

The mining engineer has to match each orebody with an appropriate mining rate. Not uncommonly, he or she is forced to conclude that mining can only realistically occur at some fraction of the maximum rate because of the awkward shape of the deposit.

In the case of an ore deposit that is oriented more or less vertically, there is a good rule-of-thumb that can be used for selecting mining rates: the daily tonnage rate should be about 15% of the number of tonnes indicated or developed per vertical metre of depth. If, for example, the exploration program has proved up 6,500 tonnes of ore per vertical metre to some reasonable depth, a daily rate of 1,000 tonnes could be justified.

The width of an orebody will have a direct bearing on production rates as well. Ore in wide orebodies can be mined and handled much less expensively per tonne than can ore in narrow occurrences.

Rate of production is also related to the ore's availability for extraction, because most mining methods require

Photo by The Northern Miner

By using very large equipment, large mining operations can be more profitable.

that the miner leave behind some ore in pillars to support the structure of the mine. It may be some time before this ore becomes available for stoping, and some of it may have to be left behind altogether.

THE RISK OF DILUTION

In some mines, the physical characteristics of the wallrock may force a company to mine a considerable amount of unwanted, barren rock along with the ore. This waste rock must then be transported to surface with the ore, so naturally, such dilution of the ore costs the mining company money. In some cases, dilution can reach as high as 20% of what is mined, making the mine far less profitable than it could have

been had the mining engineer been able to devise a method of mining only the mineralized rock.

METALLURGICAL RISKS

Some of the most successful mines now operating were commercially worthless until serious metallurgical problems could be overcome.

Other mines' ores contain metals that cannot be recovered until the ores are ground very fine or, in some cases, oxidized to liberate the valuable metals. The costs of grinding rise sharply with fineness of grind, and oxidation of small amounts of ore can be fairly expensive.

The role of the metallurgist is to select the best and most affordable process available for optimum recovery from a given ore, normally by testing drill core, pit or underground bulk samples. Care must be taken to ensure the material tested (and thus, counted on to represent the entire orebody) is fresh enough not to have oxidized or altered in any way and that there is enough of it to make a reasonable judgment of the orebody by extrapolation.

Still, even if the metallurgist is given up to thousands of tonnes of material from test mining, the sample may not be representative of the deposit. It could be from a high-grade zone, or from an area with poor ground conditions under study by the mining engineer for the risk of excessive dilution. It could also be from the top of an open-pit deposit that has been weathered more than what lies further below.

To do a good job, the metallurgist must be provided with samples representing minable grades of each geologically different part of the deposit.

The results of metallurgical tests will normally show the recovery — the fraction of metal in the feed that is recovered in the metallurgical process. If a concentration process is being tested, the results will show the concentrate grade — the amount of metal in the concentrate — and the amounts of any trace elements (such as mercury, arsenic or antimony) that are difficult for a smelter to handle or environmentally hazardous.

Feasibility Models

Once the technical and economic risks of developing a mineral deposit have been assessed, and costs and revenues estimated, the feasibility study calculates the return the project will earn on its capital cost.

The conventional method is the "discounted cash flow" model, which calculates a net present value for the project by adding up its annual cash flows, discounted so that they represent current dollars. The net present value is the "worth" of the project, the amount of cash that the project will generate over its life.

A second figure, the internal rate of return (IRR) — defined as the discount rate at which the net present value falls to zero — is taken to represent the return the capital that goes into the project will earn for its owner. Mining companies will decide to put a project into production only if its IRR exceeds a set amount.

Feasibility studies in their most definitive form are done to make a final production decision on a project and to provide potential lenders or equity partners with a financial analysis they can use to make their investment decisions.

But at earlier stages, less rigorous economic analyses — called "pre-feasibility", "scoping" or "preliminary economic assessment" studies — let exploration and mining companies assess advanced projects and decide whether to spend more money on development.

12. Metal Markets

Cashing In

As in any other industry, mineral producers have to find a market for their products. Only then can they profit from their efforts.

Some metals have an established terminal market – an institution where metal dealers bid for quantities of metal, and daily establish a price for it. Other metals are priced by individual producers, who respond to the market for the metal by raising the price in times of high demand, and lowering it when demand falls off.

For base metals, the most important international terminal market is the London Metal Exchange (LME) in London, England. The LME's member firms specialize in metals-brokering services, facilitating the purchase and sale of metals to LME specifications by clients. Metals bought and sold at the exchange include copper, lead, zinc, aluminum, tin and nickel by way of twice-daily bidding rings.

The LME functions as a clearing market for metals, offering both spot and futures contracts. Its primary role is to provide a service to clients seeking to fix prices, hedge inventories and generally manage risk from volatile commodity prices.

A buyer of futures purchases a contract for so many tonnes of metal, even though that metal has not yet been mined or processed. The contract specifies the metal will be delivered at some time in the future, usually in three months. The price offered will be different, and usually higher, than the spot price. Such futures and the warrants representing them are negotiable in the same way as spot metals.

The same reasoning underlies the trading of futures. That is, buyers seek protection against rising prices by hedging against current sales. Sellers of futures contracts, usually metal producers, ensure that their metal can be sold at the contract price, thereby "locking in" their profits.

Quotations from the LME are the basis for transaction prices in much of the world. The quotations represent actual day-to-day purchases between buyers and sellers, and they are released to news services and published in the daily press.

In North America, the Commodity Exchange (COMEX) in New York is a major terminal market in several metals. It tends to be a speculative market, attracting a much higher proportion of individual investors than does the LME. In London, the primary business is with trade accounts using terminal markets for business purposes.

Futures exchanges are basically hedge markets. A very small tonnage of the total sales on the hedge markets ever actually changes hands, physically. Most of the sales for physical supplies (metal producers selling to manufacturers of metal-based products) are handled by metal traders and/or producers. The merchant metal price generally fluctuates daily in line with movements on the LME and U.S. commodity exchanges.

The London Metal Exchange maintains a web site with recent price and trading data, at www.lme.co.uk. COMEX prices are available at www.nymex.com.

Producers have a fixed price that is set periodically in response to prices on the terminal markets. These are called producer prices. It is also common for producers to sell their metal at a fixed premium to the daily settlement prices on the LME or COMEX.

FUTURES DEALING

In a normal market, the forward quotation is usually higher than the spot price. This is because securing copper today for delivery in three months requires someone to finance the copper. The cost of the forward quotation reflects this.

When the forward quotation is higher than the spot price, the mar-

Photo by The Northern Miner

This mill in Nevada's Carlin trend produces gold that is later sold on world markets.

Photo courtesy of World Gold Council

Gold is bought and sold as bullion; the bars are stamped to certify .9999 fineness.

ket is said to be in contango. During periods of very high demand with very tight supplies, the spot quotation can rise above the forward quotation. This situation is called a backwardation. It signifies very high spot demand and the inability to supply it without offering a premium to draw it to the market.

GOLD – AS GOOD AS CASH

The price of gold is influenced by monetary, economic and political factors. For many years, until the early 1930s, its price was controlled by governments and pegged at US$20.67 per troy ounce. All gold produced in Canada was sold to the Royal Canadian Mint.

In 1934, U.S. President Franklin Roosevelt officially raised the price of gold to US$35.00 per troy ounce and, in effect, re-established the gold standard which had been displaced by floating exchange rates following the First World War.

In 1947, the Bretton Woods agreement ushered in an era of fixed exchange rates whereby various world currencies were exchangeable into the U.S. dollar, which, in turn, was readily exchangeable into gold.

This system worked well into the late 1960s, when speculative pressure against the American dollar caused a run on gold. This brought in the "two-tier" gold system, where there was an official market for central banks and a "free" market for others.

Speculative pressures and a faltering U.S. economy forced the government to raise the official price to US$38.00 per troy ounce in 1972 and again to US$42.22 the

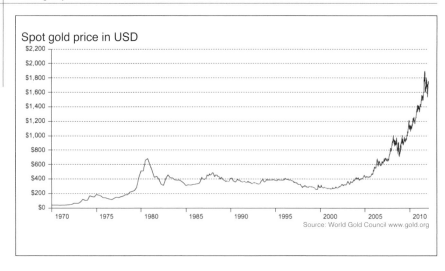

Spot gold price in USD

Source: World Gold Council www.gold.org

following year – in effect, devaluing the U.S. dollar.

Since 1972, gold has been freely traded on terminal markets. Both Zurich and London bullion markets vied for dominant influence. The Winnipeg Commodity Exchange started the trade in gold futures in 1972. COMEX and other U.S. markets followed suit to create a lively spot and futures market. Gold prices today fluctuate, based on terminal market buying and selling.

Because it is an investment of last resort, gold also functions as a currency and tends to increase in value as other currencies fall. Its price in a given currency will go up during times the currency is weak, and fall when the currency is strong.

Prices quoted on spot and futures markets are for "0.9999 fine" gold, that is, gold with a minimum 99.99% purity.

OTHER PRECIOUS METALS

The price pattern for silver is complex. Fear of inflation, armed conflict and the changing patterns of industrial usage all have effects on the price of silver. The metal is quoted in U.S. dollars per troy ounce. The dominant market for silver is the London bullion dealers' market, and there is an important futures market at the Comex in New York. Prices quoted on spot and futures markets are for 0.999 silver.

The prices of platinum and palladium are fixed daily by a group of dealers in London, and futures are traded on the New York Mercantile Exchange. Prices are quoted in U.S. dollars per troy ounce.

Other precious metals of the platinum group do not trade on terminal markets, and the most useful quotations are the producer prices set by individual refiners in response to the market for each metal. The precious-metal dealers in London belong to the London Bullion Market Association, which maintains www.lbma.org.uk with recent market statistics, including the daily prices. Montreal-based precious metal dealer Kitco has a useful market snapshot at www.kitco.com, with prices for gold, silver, platinum, palladium and rhodium. The World Gold Council's www.gold.org is also handy.

OTHER METALS

Prices of metals that are not traded on terminal markets like the LME or the bullion dealers' market generally find their price levels by supply and demand. Many are traded on long-term contracts between consumers and producers – for example, a steel producer might have contracts for the supply of iron ore, chromium, and nickel with several producers (i.e. miners).

In the absence of daily spot prices and futures prices, the important price is the producer price. Most often producer prices will be set for several different grades or forms of the metal. For example, there are separate prices for refined cobalt, cobalt powder and cobalt oxide.

Similarly, iron ores are graded for sale, based on the amount of contained phosphorus, silica and other impurities. Iron ore pellets are the preferred form of iron among steelmakers. Prices are quoted per tonne for international trade.

Metals used in steelmaking, such as manganese, chromium and vanadium, may be sold as refined metal, ore concentrate or alloyed – at a specified concentration – with iron.

Two other important alloy metals without terminal markets are tungsten and molybdenum. Tungsten is sold as ore concentrate or ferrotungsten, while molybdenum is generally quoted at a price per pound of molybdenum contained in a molybdenite concentrate.

The price of uranium is affected by political factors related to its military use. While a market does exist for sales of small lots on a spot basis at a price called the "exchange value", most uranium is sold to public utilities under long-term contract.

No matter how the metal occurs in the mine, uranium is sold in U.S. dollars per pound of U_3O_8 (a uranium oxide). Mine output is priced accordingly.

DIAMONDS AND THE GEMSTONES

What sets diamonds and other gems apart from metals is the fact that the value of a company's production over a given period of time depends not only on volume, but on quality. In other words, gold is gold and copper is copper, but there are over 14,000 price categories of rough diamonds. The quality and, thus the value, of a diamond is based on four parameters known as "the Four Cs": crystal shape, color, clarity and carat weight.

Diamond projects are evaluated not only by the average dollar value per carat of the diamonds produced, but also by the grade (carat per tonne) and tonnage in a kimberlite or lamproite, the two most common host rocks for diamonds. These factors vary from mine to mine. The large Argyle mine in Australia has high grades, but the value per carat is low at well below US$20.

Canada's first diamond mine – Ekati in the Northwest Territories – has compared favourably with the world's best mines in terms of dollar value per carat (an average of US$85 per carat) and grade (1.09 carats per tonne), but the tonnage has been generally lower than in some other kimberlites. To compensate, Ekati has exploited multiple kimberlite pipes, rather than one. It has produced 3.5 million to 4.5 million carats of diamonds per year, about 6% of current global production by value. Canadian production reached 12% to 15% of global production as several mines nationwide were developed.

Diamonds make their way from mine to jewelry store via the "diamond pipeline," which is still largely controlled by the world's most famous and successful monopoly: De Beers of South Africa. De Beers set up its Central Selling Organisation (CSO) in 1930 and, for the next seven decades, the CSO promoted the sale of diamonds from De Beers' own mines and from associated producers. Even up to the mid-1990s, the CSO was involved in the sale of more than 80% of the world's diamonds.

However, in the past decade the construction of several new diamond mines in Australia and Canada by De Beers' competitors has loos-ened the CSO's monopolistic grip. De Beers has responded by changing the name of the CSO to the "Diamond Trading Company," and by changing the focus of the organization towards marketing and retailing De Beers diamonds specifically rather than all diamonds generally.

The turn of the millennium also brought with it the issue of "conflict" or "blood" diamonds — that is, diamonds that are mined in civil or tribal war zones and, by their sale, allegedly help accelerate and prolong these conflicts, especially in Angola, Sierra Leone and the Democratic Republic of the Congo.

Not wanting their luxury product tainted, diamond producers have adopted measures to identify the point of origin of diamonds, and have pledged not to buy diamonds from artisanal miners and diamond brokers in war zones.

Whether or not those lofty goals are attainable, the concern over conflict diamonds has benefited the Canadian diamond industry, with its strict auditing of all diamonds produced, paucity of independent rough diamond dealers, and great distance from the world's war zones.

13. Making Sense of the Numbers

UNDERSTANDING FINANCIAL STATEMENTS

Those long columns of numbers at the back of a company's annual and quarterly reports may look dry, but they can reward the investor who takes the trouble to give them careful study. By learning the past earnings record and current financial health of a company, an investor can decide whether or not a stock is worth investing in. The better informed the shareholder, the safer the investment is, especially with high-risk junior companies.

Simply put, the annual and quarterly reports are the formal account of the past term's financial activities and operations. Not only do such reports have to meet acceptable accounting standards, they must contain information required under national, provincial or state securities legislation and corporation laws. In addition, shareholders must receive them within a specified period of time.

Annual reports show the names of directors and details of remuneration for officers and directors, including details of their share purchase plans; a list of investments in other corporations wholly or partly owned by the company; details of long-term debts; and information on lawsuits the company may be facing.

THE BALANCE SHEET

The balance sheet illustrates the financial picture of the company at a specific date, usually the closing day of the company's financial year. Included are the corporation's assets, liabilities and shareholders' equity.

On one side of the balance sheet are listed the company's assets, which are anything that the company owns or has owing to it. They include the current assets, such as cash, short-term securities, accounts receivable, inventories and prepaid taxes, and the fixed assets, such as buildings, factories, machinery and equipment.

Note that fixed assets are sometimes referred to as "property, plant

Northern Miner photo

Out of the vault: scenes like this one, from the old trading floor of the Toronto Stock Exchange, have passed into history as exchanges have developed "virtual" trading floors that allow traders to match orders from their own computer screens.

and equipment." The normal method for valuation of these fixed assets is the cost minus whatever depreciation has accumulated by the date of the balance sheet. Accountants regard depreciation as the decline in useful value of a fixed asset due to wear and tear from use over time.

The other side of the balance sheet shows liabilities and shareholders' equity, or the net worth of the company, which represents the shareholders' interest.

Liabilities are what the company owes others or, more specifically, all debts that fall due in the coming year and beyond. Included here are current liabilities, such as accounts payable, income tax payable and the amount of long-term debt paid off that year, and long-term liabilities, which are debts due one year or more after the date of the financial report.

Long-term .liabilities include not only long-term debts, but also deferred income taxes. This is income tax that would otherwise be payable, but which is deferred by using certain deductions provided by the government. Any tax writeoffs in the early years of investment serve to reduce what the company would otherwise owe in current taxes.

Shareholders' equity is the total equity interest that all stockholders have in the company. On the balance sheet, liabilities are subtracted from assets, and what remains is shareholders' equity or ownership in the company. In this way, assets always equal liabilities plus shareholders' equity. For legal and accounting reasons, it is usually separated into three categories:

• Capital stock, which are the shares representing ownership of the business, including preferred and common;

• Retained earnings, which are the after-tax profits over the life of the company after all expenses and dividends have been paid out; and

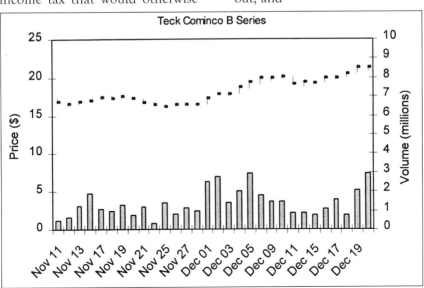

Stock charts show prices (upper lines) and volume of shares traded (vertical bars).

• Contributed surplus (sometimes called capital surplus), which is the amount raised by the sale of shares in excess of the par or market value of each share.

The balance sheet does not show how much revenue a company took in during the year. Nor does the balance sheet show the expenses incurred, how much profit was earned or loss incurred. This information is provided in the earnings statement.

THE EARNINGS STATEMENT

Sometimes called the income statement or profit-and-loss statement, the earnings statement shows how much money a company made or lost during the year from the sale of products or services, and the expenses the company incurred for wages, operating costs, etc. The difference between how much was taken in and how much was spent is the net earnings, or profit, of the company. This money is used to re-invest in the company and to pay dividends to shareholders.

The earnings statement is divided into four main sections: the operating section, the non-operating section, the creditors' section, and the owners' section.

The operating section lists income from the sale of the company's goods or services, minus the cost of sales (labour, energy, etc.) This provides the gross operating profit.

In the non-operating section, non-operating income, such as interest and dividends from company investments, is added to the net operating profit. To this is added "extraordinary items", which are any unusual and significant additions to income or losses (the one-time sale of a large asset, for example). The sum of these represents the company's remaining income from all sources.

Payments to creditors are listed in the creditors' section. These usually take the form of fixed interest charges on bank loans and interest charges to other debt-holders who have lent money to the company. These payments are deducted from the income of the company.

Finally, we have the owners' section, which shows the company's net earnings or deficit. The net earnings are shifted to the retained earnings statement, which shows the total annual earnings retained after payment of all expenses and dividends.

THE RETAINED EARNINGS STATEMENT

This portion of the annual report shows the amount of earnings which have been kept in the business, either as cash or reinvested in new assets. Stated another way, it reveals the excess of net earnings that have been accumulated by a company year-by-year, over and above dividends paid out to shareholders.

CHANGES IN FINANCIAL POSITION

This section of the report explains changes in working capital (current assets minus current liabilities) between two consecutive years. It goes by many names (including "Source and Application of Funds Statement" and "Source and Use of Funds Statement"), but basically acts as a bridge between a company's balance sheet for those two years. It also summarizes how a company

Photo by The Northern Miner

This mine in Nevada produces gold and earnings for Barrick Gold.

raised financings for the year and how those financings were used.

AUDITOR'S REPORT

Companies are legally required to appoint an outside, independent accountant to represent shareholders and report to them each year on the company's business. Essentially, the accountant's job is to verify the accuracy of the company's financial statements. The auditor's report is usually quite brief, and is meant to give the shareholder assurance that the annual report is a reliable indication of the company's financial health. Unfortunately, in recent years, the brief report came with an equally cursory examination of the company's finances — and the resulting financial scandals in the 1990s and 2000s led some investors to question the competence and professionalism of the major auditing firms. The accounting profession, fortunately, took steps to repair the damage done by poor auditing practices, and with financial accounting practices gradually being standardized around the world, the auditor's letter can again be seen as a measure of certainty in financial reporting.

NOTES TO THE FINANCIAL STATEMENT

Finally, interesting information is often found in the notes to the financial statements in annual reports. Study these carefully, as they will sometimes disclose details of lawsuits filed against the company, or other information useful to investors.

14. Investing in Mining

No Substitute for Common Sense

The common-sense rules of investing are as true in mining speculation as they are for investment in any other industry. If a magic formula existed to identify the mining ventures that would prove to be spectacular successes, every speculator would have retired on his riches long ago. An earlier author of this book closed a discussion of investing with the words, "above all else, investigate before you invest." It was true then and it is true today.

Investing in large mining companies that have a number of mines and produce a number of commodities is much the same as investing in other corporations like manufacturing, utility, or banking firms. Securities law in the countries where their stocks trade compel the companies to make their financial records public; analysts, investment dealers and the financial press study the quarterly and annual results and offer their opinions. The investor can use these professional opinions to form his own picture of a company's prospects.

Anyone considering investing in the established mining companies, or in large established firms in any industry, should also consider whether the stock market as a whole is a good investment at the time. Many individual investors buy into the stock market when it is doing well, during a period of economic growth. It is during those times, however, that the market places a high value on shares. Buy then and you will pay a high price for stock. This creates the risk of being stuck with it or selling at a loss when the economy recedes and stock prices fall.

Investors laugh when they are told the old rule, "buy low and sell high", as though the advice is so obvious it needn't be said. Then they go online and buy a stock that has just traded at its all-time high on record volume. An investor will often get better results by buying shares when the market has declined and prices are low. It is also wise to seek out stocks that are not yet widely recommended – after all, if everyone has already bought Acme Widgets, there will be no one left to buy the stock and force its price higher. Forget Acme, and look for a better buy elsewhere.

JUNIOR MINING COMPANIES

Investing in the "juniors", the small companies formed to explore for mineral deposits, is much more speculative than investing in larger companies. Junior exploration companies, as well as other venture-capital companies like small high-technology firms and junior oil explorers, offer high returns if they are successful. The other side of the same coin is the higher risk that they will fail.

One very sensible rule is never to invest money you cannot afford to lose. Understand that the junior exploration game is a gamble, and keep the rent money out of it. The buy-low/sell-high bromide means more here than anywhere else in the market for once the stock's price is up and everybody is talking about a great mining play at Dead Cat Falls, the easy money has probably been made.

Jumping from one stock to another only makes money for your broker. Picking a good investment and staying with it is often a better strategy. One way of picking that stock is to look at its people. If they have a record of making projects go ahead, it's a good sign that they are diligent and talented.

Beware of promotions. In particular, never buy a stock from a dealer who calls you up cold – you haven't had time to investigate the company's management or assets. High-pressure investment sales people are still around. Dealing with them will cost you money.

Be extremely wary of information from Internet chat rooms and bulletin boards. Some of the people posting bullish statements are long on the stock, while some of the negative messages may originate from professional short-sellers. The exchange of views may be entertaining, but they should not be relied upon to make sound investment decisions.

Remember, too, that there is little or no regulatory oversight of newsletter writers, and many hold positions and are on the payroll of the companies they are recommending. While some news-letters provide quality information, others are purely promotional. Stick with experienced, respected newsletter writers and avoid the rest.

If you have an opportunity to invest in a prospecting syndicate, keep in mind that these investment rules are comparable to those for investing in junior exploration companies. Invest only what you can afford to lose. Remember that prospecting is a gamble. Consider the management carefully; is there an experienced prospector or mining professional who can pilot the venture successfully?

Recall that the success of a prospecting venture depends on getting an exploration or mining company interested in the property. This is much easier to accomplish during good times; companies will never have money to burn, but they are

Photo courtesy of the Prospectors and Developers Association of Canada

A great place to meet the management of junior mining companies is at trade shows, such as the Prospectors and Developers Association of Canada convention held in Toronto every March.

more likely to have money and manpower to commit to a good-quality prospect when they are making money themselves.

DISCLOSURE

A public company is required, under the relevant securities regulations, to make "full, true and plain disclosure of all material facts" in its prospectus when it goes public or otherwise issues shares in the primary market and, thereafter, to make continuous financial disclosure and timely disclosure of material changes. Junior companies often have no revenue, rendering their financial disclosure of limited relevance. However, their technical disclosure filings can and should be used by investors to make informed decisions.

Photo by The Northern Miner

A salting scandal at Bre-X Minerals' Busang project in Indonesia led to tighter rules for junior exploration companies. Shown above are members of Bre-X's geological team.

A mining company's disclosure obligations begin with its activities (such as the way in which it conducts its exploration program), through the assembly of technical information (such as drilling records and assay results), to the production of technical filings, filing such reports with regulatory authorities and the disclosure of the information by way of public news release.

It is here that the Internet becomes a valuable tool for the investor. Securities regulators or stock exchanges around the world now operate websites that compile and present the news releases, quarterly and annual reports, and other documents that companies are required by law to file. In Canada, the provincial securities regulators have set up the System for Electronic Document Analysis and Retrieval, accessible at www.sedar.com. The Securities and Exchange Commission in the United States has a similar clearinghouse (called EDGAR, for "Electronic Data Gathering, Analysis, and Retrieval") at www.sec.gov.

In the United Kingdom, the London Stock Exchange puts corporate news filings on its website, and news from Australian-listed companies is posted by the Australian Stock Exchange.

The web archives of news dissemination services, and the websites of the companies themselves, are also good places to look for current information and filings.

Analysts' interpretation for clients provide a different kind of disclosure about mining companies. A number of parties are involved in the process, including mining companies, promoters, brokers and analysts and investors themselves.

Several exchanges now require that a "Qualified Person" be responsible for designing programs and ensuring that these programs are carried out in accordance with industry standards. Any failure by a Qualified Person in the discharge of these responsibilities would make the QP subject to disciplinary action from his or her professional organization and, possibly, regulatory authorities. In addition, officers and directors who fail to discharge their duties could face limited statutory liability for the misleading continuous disclosure of their public companies.

Brokers and mining analysts are regulated by professional associations and securities regulators. In some jurisdictions, new regulations oblige them to differentiate between their opinions and information provided by mining companies, as well as provide the basis for any conclusions, opinions or calculations contained in their reports.

Under new guidelines imposed by most Canadian exchanges, technical reports on exploration results must include:

- results of all surveys and investigations on the property;
- details of the interpretation of exploration information and plans for future work on the property;

- a description of the geology, mineral occurrence and nature of mineralization found:
- mineral distribution, rock types, structural controls, the cutting criteria used to establish the sampling interval and the identification of any significantly higher-grade section within a lower-grade interval;
- details of the location, number, type, nature, spacing or density of samples collected and the area covered;
- details of the type of assaying or analytical procedure used, sample size and name and location of the assay or analytical laboratory used and its accreditation; and
- the true width of the individual assays, to the extent known.

Interpreting Information

All the disclosure in the world will not help investors if they do not know how to interpret technical information. It is important to realize that each mineral project is different, and what is relevant to one project may be of no significance to another. It is also important to realize that mining companies promote their projects by presenting them in the best possible light. As a consequence, investors must learn to interpret the technical information contained in news releases, such as drill results and reserve and resource calculations, in order to understand their economic significance.

Drill Results

Drill rigs are not called "truth machines" for nothing. No amount of promotion or favourable press will prevent companies from being affected by negative assay results when good ones were anticipated. A selloff takes place immediately, with sophisticated players leading the way. That is why it is important for investors to understand the economic significance of drilling results, relative to the nature of the deposit being explored and where it is situated. You do not want to be the one buying when better-informed investors are selling, and vice versa.

The first step to analyzing these results is to understand what grades will be economic in a typical mining operation. In general, large deposits capable of being mined by open pit will have lower cutoff grades than will deposits of the same metal that have to be mined by underground methods. For example, the Dome gold mine in northern Ontario operated an open pit with reserves grading 2.3 grams per tonne; its underground reserves graded 4.3 grams.

A 30-metre drill intersection grading 1 gram gold per tonne, starting at 5 metres below surface, would be of economic interest if it originated from a road-accessible property in Nevada. The state has numerous low-grade deposits mined by less costly open-pit methods, whose gold is recovered by low-cost heap leaching. However, this same drill intersection would not attract much interest on a property in the remote Arctic owing to a lack of roads and other infrastructure.

A drill intersection showing 0.4%

copper would be exciting in the porphyry-copper districts of Chile, since it approaches the typical grade of the area's large open-pit mines. In the volcanic belts of northern Ontario and Quebec, where smaller, higher-grade base metal mines are the norm, the same result would merely merit more work.

Deposits with complex metallurgy will be more costly to mill, so cut-off grades rise accordingly. Off-site costs such as transportation are a major component of total production expenses. This is particularly true for gold and base metal deposits that produce concentrates, which must be shipped to a smelter for processing. Energy costs have a strong influence on a potential operation's economics, especially when a mill must be built. As a general rule, a mineral deposit in an area where access is poor and costs are high will be less likely to make a mine than an equivalent deposit in a more developed region.

The length of drill intersections is also important, since these give an indication of the kind of mineralization present. Low-grade deposits must be large enough for bulk mining. That means low-grade drill intersections must be long – on the order of tens to hundreds of metres – to show economic potential.

Narrow intersections, on the other hand, must show high grades. A zone's width is vital to the feasibility of mining operations, and narrow zones may simply not be economic to mine. A mineralized zone only a fraction of a metre wide could not be mined without also extracting large amounts of wall rock, thereby diluting the grade.

Core lengths quoted in press releases can also be misleading. Drill holes that cross a zone of mineralization at nearly a right angle give a better picture of the zone's true width. Holes that are drilled at shallower angles to the mineralized zone will yield intersections with lengths much greater than the true width of the mineralization.

Keep in mind that a hole drilled to a great depth from surface will often be drilled at a very steep angle, such that if it intersects nearly vertical structures at depth, the true width will be much less than the length of the drill intersection.

There is always the possibility that the company has drilled down the dip or plunge of a mineralized zone rather than across it. This can happen, on occasion, in the early stages of an exploration project when the structure being explored is still poorly understood. A hole drilled in this way will provide no information about the width of the zone, yet the mineralized intersections may run to spectacular (and meaningless) lengths.

Lastly, it is important to consider that narrow zones of high-grade mineralization may "carry" an intersection. A zone only a metre wide may show a grade of 15% zinc, but an over-promotional company may aver-

Photo by The Northern Miner

Investors need to keep in mind the security risks of operating in developing countries.

Photo by The Northern Miner

Prospector Albert Chislett shows drill core from his Voisey's Bay nickel discovery in Labrador.

age that one-metre intersection with the nearly barren two metres to each side of it, reporting five metres grading 3% zinc. This is sometimes called "rubberholing."

In summary, it is important to consider both grade and width when assessing exploration results, and to remember that drill results obtained at one property are not comparable to results from another, unless both properties are in the same geological region and have similar geological and metallurgical characteristics. To assess a drill result, consider the typical mining grade of similar deposits in the same area.

UNDERSTANDING RESERVES AND RESOURCES

As noted elsewhere in this book, a resource is a body of mineralization in the ground, whereas a reserve is that part of the resource which is economically minable. It is not unusual to find that no part of the deposit is economically minable. The two should not be confused but, too frequently, they are. Investors, writers, analysts and promoters that lack knowledge, or are frugal with the truth, may describe a mineral deposit as a reserve even though its minability has not been demonstrated. This often leads to unrealistic values being placed on projects, on the companies that own them, and on the shares of those companies in the market.

Economic planning for a mine can only be based on minable reserves. Banks will normally only lend money to develop mineral properties to companies that have completed a full feasibility study of their project, based on proven and probable reserves. The study has to include a mine plan or pit design outlining exactly what fraction of the body can be mined, and also must include the results of metallurgical tests showing how much metal can be recovered from the rock.

Moreover, the lender will insist that the study be done by a recognized independent consulting firm, not by the company itself, and may seek other opinions before lending the money.

Investors should be sure they understand whether a company is reporting reserves or resources, and should also know how certain an estimate has been made – whether the reserve is proven or probable and a resource measured, indicated or inferred.

Once reserve and resource estimates have been made, comparisons with operating mines are often instructive. Reserve figures for mines are often published in reference works like the *Canadian and American Mines Handbook*, and can also be found in the annual reports of operating companies.

RED FLAGS AND SCAMS

Mining is a respectable business, but lurking in the shadows are silver-tongued scoundrels and pseudo-scientists who profit from dishonest promotions and deceitful scams. They were around in Mark Twain's time and they are still around today. The infamous Bre-X Minerals salting scam of the mid-1990s was among the most boldly executed swindles in mining history.

Most swindles involve gold, which has a mystique and allure that attracts not only charlatans, but enthusiastic investors anxious to believe that riches are at hand. The scams range from stock manipulation and misleading announcements to fraudulent assays and sample-tampering or "salting".

The classic stock manipulation is the "pump and dump." An obscure stock, usually trading at only pennies per share, is suddenly the subject of favourable information, brokers' recommendations and street talk that fuels demand and drives up price. The company's management, and sometimes other shareholders, take advantage of this hype to sell large amounts of stock acquired earlier at a lower price. The news then dries up, the price collapses and small investors who bought on the favourable news are left with shares acquired at the high end of the market.

In some cases, the pump and dump is often done in collusion with a securities firm that holds enough stock to act as a market-maker in the company's shares.

There are other manipulative techniques, such as "high-closing," in which two holders make a trade at a higher price near the closing bell of the market, creating the illusion of price movement. Large block trades between company insiders or market-making securities firms can likewise serve to create the illusion of trading volume.

In other schemes, devious promoters sometimes show investors spectacular samples from old gold mines containing visible gold. Be wary of high-grade assays from narrow quartz veins. While some veins can be economically exploited, many are uneconomic because they are too narrow to be mined using modern techniques. Remember, a few high-grade results do not a mine make.

SALTING

Salting scams are few and far between, and are usually detected quickly because assay results are abnormally high or do not conform to what might be expected from the geological setting of the region. However, the dubious art is always being refined, and sophisticated methods have been used in some recent salting swindles, including that of Bre-X Minerals at its Busang property in Indonesia.

Salting is the deliberate introduction of metal into a sample, intended to produce a false assay result. There is a long and ignoble history of salting scams in mineral exploration, usually in precious metal exploration, which lends itself to sample tampering. The small amount of gold, silver or platinum needed to affect the result of an assay makes the practice relatively easy. Also, because genuine precious metal mineralization is often not visible to the naked eye, barren and mineralized samples may look similar, making it harder to detect salting.

Often coarse, placer gold is added to barren samples, which results in erratic assay results and so-called "reproducibility problems" far beyond what might be expected from a naturally occurring deposit with coarse gold. The salting is usually detected through testwork which shows that all of the gold found in the samples is coarse-grained, with little or no fine grains. Normally, if coarse gold occurs in a deposit, there is always plenty of fine gold along with it.

Salting scams are not as common in base metal exploration, though they are not unknown. One reason is that base metal mineralization is usually visible in drill cores and hand specimens, and it is easy to recognize salted barren samples. There may be another reason: base metals just aren't quite as romantic as precious metals, and don't stir the imagination of naive investors. However, even in recent years, the share prices of several juniors have soared to great heights based on visual estimates of the base metal content in drill core. "Eyeball assays" are now frowned upon by securities regulators because, in

Photo by The Northern Miner

Don't be fooled by fancy-looking equipment at an exploration site. Fabulous assay results from the property shown above were later found to have come from salted samples.

most cases, the subsequent assays show little or no mineral content, but only a need for the geologists to invest in corrective eyewear.

COOKED ASSAYS

In his book Roughing It, Mark Twain tells the story of a popular assayer who never failed to get spectacular results from all mineral specimens brought to him for testing. The town's respectable citizens became concerned and conspired to send him a fragment from a carpenter's grindstone for assaying. When the fabulously rich results were published in a newspaper expose, the popular "assayer" was forced to leave town in great haste.

Phony assay results are as old as the hills, and a favourite ploy of the crook is to point to an "assay discrepancy" once confronted with an independent evaluation of the property he is promoting.

In 1987, the Northwest Mining Association warned that US$250 million was lost that year alone to "dirt-pile swindles", gold scams based on phony assay results. Sometimes called "desert dirts", these kinds of swindles often prey on wealthy retirees with little or no knowledge of mineral exploration.

As in Mark Twain's time, certain laboratories produce data showing exceptionally high concentrations of gold, and sometimes platinum group metals, in almost every rock sent to them for assaying. Mining professionals never use these laboratories because of their dismal reputations.

Fraud artists have found a number of ways to falsify the results of an assay. Samples themselves may be salted, or an accomplice placed in the assay laboratory to add metal to the chemicals used in the analytical process. Numbers may be falsified or assay certificates counterfeited.

Closely related to the salting scam is the assertion that a deposit contains "unassayable" or "refractory" precious metals that can only be detected using a "proprietary" technique. This is the realm of quack science, in which con men and deluded pretenders talk the language of chemistry and metallurgy without regard for the scientific method.

The con artists use any number of excuses to explain why conventional chemical techniques cannot find their gold. Some say that their gold is "micro-fine" or in "micro-clusters" that take on chemical characteristics that prevent them from being assayed. This explanation flies in the face of overwhelming evidence that when the particles of a substance are more finely divided,

Photo by The Northern Miner

The "millions of ounces" at the Busang project were later found to be a hoax.

it will react more readily with other substances and be extracted more easily from its matrix.

Other unassayable-gold promoters declaim about "encapsulation" by other minerals, usually silica, arguing that the surrounding minerals seal off the gold from assay fluxes and leaching acids. Yet there are dozens of strong acid attacks and reactive fluxes that will destroy any "encapsulating" minerals. A third scapegoat is "interference" by other elements, which is alleged to prevent analytical instruments from detecting the metals of interest. The problem is well-known in analytical chemistry and a capable chemist can find ways to eliminate the interfering elements.

In short, there is always a legitimate and reasonable chemical argument that can be made against a claim of "unassayable" gold. There are also a number of analytical methods, such as neutron-activation analysis, that are not sensitive to the supposed causes of unassayable metal. Nevertheless, the investor will usually be a layperson unacquainted with analytical chemistry and must, instead, keep eyes open for red flags and other signals that a project is far from legitimate.

In the early 1980s, the Canadian Institute of Mining, Metallurgy and Petroleum published a paper debunking the myth of "unassayable" deposits. The authors investigated many of these projects and found that in every case, the fabulous numbers reported from unconventional assaying methods were the result of technical incompetence or outright fraud.

DANGER SIGNALS

Some danger signals are obvious. The old saw that if something sounds too good to be true, it probably is, will likely be reliable until the end of time. There are other signs that a play is more promotion than prospect, and some are listed below. Doubtless other red flags will be identified by future generations, for con artists can be very original.

- Any claim that samples require a "proprietary" analytical method: this is nonsense.
- Unusual emphasis on sample security or "chain of custody" before analysis: the management doth protest too much.
- Any failure to split core and retain half for independent checks: the destruction of evidence.
- "Verifications" by independent consultants that are no verification at all: for example, when a consultant acknowledges that samples contain precious metals, but does not verify actual grades.
- Reserve or resource estimates based on surface sampling or trenches, with little or no evidence from drill holes at depth: no responsible geologist would suggest it is possible to do that.
- Reserve or resource estimates based on very widely spaced drill holes or limited drill information: nature reserves the right to spring a surprise on anyone who digs a large hole in the ground.
- Suggestions that a deposit is so large that entering into production could substantially affect the market for that metal or "put the big producers out of business."

that is invariably pure hype.
- Dark mutterings of a "conspiracy" to prevent technological progress or to keep the project out of production; similarly, mentions of a "conspiracy" by short-sellers in the market: conspiracy theories are standard among pseudo-scientists.
- Large cash payments made to buy the property, often from management itself: other people's money, paid up front, on a speculative venture.
- Poorly documented land tenure: if the company takes poor care of its property holdings, the property itself cannot mean much to management.

- Location of a project in a very remote or politically difficult jurisdiction where there has been little modern mineral exploration, and past efforts were fruitless.
- Personal communication with the company consistantly paints a far rosier picture of the project than the company's public disclosure.

Finally, when in doubt, ask for an outside opinion. Rather than learn these and other lessons the hard way with your hard-earned money, seek the advice of mining professionals and respected mining analysts, or government geologists and mining organizations. Learn to do your own due diligence.

Glossary

A

Acidic precipitation – Snow and rain that have a low pH, caused by sulphur dioxide and nitric oxide gases from industrial activity released into the atmosphere.

Acidic rocks – Igneous rock carrying a high (greater than 65%) proportion of silica.

Acid mine drainage – Acidic run-off water from mine waste dumps and mill tailings ponds containing sulphide minerals. Also refers to ground water pumped to surface from mines.

Adit – An opening driven horizontally into the side of a mountain or hill to provide access to a mineral deposit.

ADR plant — A hydrometallurgical plant housing circuits for adsorption, desorption, and refining.

Aerial magnetometer – An instrument used to measure magnetic field strength from an airplane.

Aeromagnetic survey – A geophysical survey using a magnetometer aboard, or towed behind, an aircraft.

Agglomerate – A breccia composed largely or entirely of rounded fragments of volcanic rocks.

Agglomeration – Cementing crushed or ground rock particles together into larger pieces, usually to make them easier to handle; used frequently in heap-leaching operation.

Agitation – In metallurgy, the act or state of being stirred or shaken mechanically, sometimes accomplished by the introduction of compressed air.

Airborne survey – A survey made from an aircraft to obtain photographs, or measure magnetic properties, radioactivity, etc.

Alloy – A compound of two or more metals.

Alluvium – Relatively recent deposits of sedimentary material laid down in river beds, flood plains, lakes, or at the base of mountain slopes. (adj. alluvial)

Alpha meter – An instrument used to measure positively charged particles emitted by radioactive materials.

Alpha ray – A positively charged particle emitted by certain radioactive materials.

Alteration – Chemical changes in minerals occurring after a mineral is formed; typical of the reaction between mineralizing fluids and host rocks, and of the surface weathering of rocks. Common types (and their characteristic minerals) include albitization (sodium feldspar), argillization (clays), chloritization (chlorite), potassic alteration (potassium feldspar and biotite), propylitization (epidote), sericitization (white mica), and silicification (quartz).

American Depository Receipt (ADR) – A security traded on United States markets, representing shares in a foreign company held by banks or trust companies in the United States; a means of trading foreign shares without those companies qualifying as share issuers under United States law.

Amorphous – A term applied to rocks or minerals that possess no definite crystal structure or form, such as amorphous carbon and obsidian.

Amortization – The gradual and systematic writing off of a balance in an account over an appropriate period.

Amphibolite – A gneiss or schist largely made up of amphibole and plagioclase minerals.

ANFO – Acronym for ammonium nitrate and fuel oil, a mixture used as a blasting agent in many mines.

Annual report – The formal financial statements and report on operations issued by a corporation to its shareholders after its fiscal year-end.

Anode – A rectangular plate of metal cast in a shape suitable for refining by the electrolytic process.

Anomaly – Any departure from the norm which may indicate the presence of mineralization in the underlying bedrock.

Anthracite – A hard, black coal containing a high percentage of fixed carbon and a low percentage of volatile matter.

Anticline – An arch or fold in layers of rock shaped like the crest of a wave.

Antiform – A fold similar to an anticline, but where relative ages of the folded rocks are unknown.

Apex – The top or terminal edge of a vein on surface or its nearest point to the surface.

Aqua Regia — A solution of three parts hydrochloric acid to one part nitric acid, one of a few solvents able to extract gold; the name (Latin for "royal water") was bestowed by Medieval alchemists.

Ash – The inorganic residue remaining after ignition of coal.

Assay – A chemical test performed on a sample of ores or minerals to determine the amount of valuable metals contained.

Assay foot (metre, inch, centimetre) – The assay value multiplied by the number of feet, metres, inches, centimetres across which the sample is taken. Used in calculating the average grade of a resource.

Assay map – Plan view of an area indicating assay values and locations of all samples taken on the property.

Assay ton — An assay charge of 29.166 grams; a charge of that weight produces a gold bead whose weight in milligrams equals the grade of the original sample in troy oz. per ton (because a short ton weighs 29,166 troy oz.). In the days before pocket calculators, this was a convenience.

Assessment work – The amount of work, specified by mining law, that must be performed each year in order to retain legal control of mining claims.

Atomic absorption spectrophotometry — An instrumental method of chemical analysis that identifies metallic atoms by the specific wavelengths of light they absorb. One of the earliest rapid and accurate instrumental methods, and still widely used.

Authorized capital – see capital stock.

Autogenous grinding – The process of grinding ore in a rotating cylinder using large pieces of the ore instead of conventional steel balls or rods.

B

Back – The ceiling or roof of an underground opening.

Backfill – Waste material used to fill the void created by mining an orebody.

Background – Minor amounts of radioactivity due not to abnormal amounts of radioactive minerals nearby, but to cosmic rays and minor residual radioactivity in the vicinity.

Back sample – Rock chips collected from the roof or back of an underground opening for the purpose of determining grade.

Backwardation – A situation when the cash or spot price of a metal stands at a premium over the price of the metal for delivery at a forward date.

Balance sheet – A formal statement of the financial position of a company on a particular day, normally presented to shareholders once a year.

Ball mill – A steel cylinder filled with steel balls into which crushed ore is fed. The ball mill is rotated, causing the balls to cascade and grind the ore.

Banded iron formation – A bedded deposit of iron minerals.

Bankable — Acceptable to lenders as a basis for financing a project; most often used to describe definitive feasibility studies.

Basalt – An extrusive volcanic rock composed primarily of plagioclase, pyroxene and some olivine.

Basal till – Unsorted glacial debris at the base of the soil column where it comes into contact with the bedrock below.

Basement rocks – The underlying or older rock mass. Often refers to rocks of Precambrian age which may be covered by younger rocks.

Base camp – Centre of operations from which exploration activity is conducted.

Base metal – Any non-precious metal (eg. copper, lead, zinc, nickel, etc.).

Basic rocks – Igneous rocks that are relatively low in silica and composed mostly of dark-colored minerals.

Batholith – A large mass of igneous rock extending to great depth with its upper portion dome-like in shape. Similar, smaller masses of igneous rocks are known as bosses or plugs.

Bauxite – A rock made up of hydrous aluminum oxides; the most common aluminum ore.

Bear market – Term used to describe market conditions when share prices are declining.

Bedding – The arrangement of sedimentary rocks in layers.

Bench testing – Early laboratory-scale metallurgical testing to determine whether metal can be extracted from a mineralized sample.

Beneficiate – To concentrate or enrich; often applied to the preparation of iron ore for smelting.

Bentonite – A clay with great ability to absorb water and which swells accordingly.

Bessemer – An iron ore with a very low phosphorus content.

Bio-leaching – A process for recovering metals from low-grade ores by dissolving them in solution, the dissolution being aided by bacterial action.

Biotite – A platy magnesium-iron mica, common in igneous rocks.

Bit – The cutting end of a drill frequently made of an extremely hard material such as industrial diamonds or tungsten carbide.

Blackjack – A miner's term for sphalerite (zinc sulphide).

Black smoker – Volcanic vent found in areas of active ocean floor spreading, through which sulphide-laden fluids escape.

Blaster – A mine employee responsible for loading, priming and detonating blastholes. **Blast furnace** – A reaction vessel in which mixed charges of oxide ores, fluxes and fuels are blown with a continuous blast of hot air and oxygen-enriched air for the chemical reduction of metals to their metallic state.

Blasthole – A drill hole in a mine that is filled with explosives in order to blast loose a quantity of rock.

Blasting agent — A relatively insensitive material capable of reacting with other chemical compounds to form an explosive mixture. Ammonium nitrate is the most common blasting agent.

BLEG — Bulk leach extractable gold; a geochemical analysis technique that takes a large quantity of soil or rock (typically around 1 kg in weight) and leaches gold from it using a solvent such as sodium cyanide solution or aqua regia. Gives a very representative result, because of the large sample size, but may not be a total extraction and may therefore understate the actual gold content.

Blister copper – A crude form of copper (assaying about 99%) produced in a smelter, which requires further refining before being used for industrial purposes.

Block caving – An inexpensive method of mining in which large blocks of ore are undercut, causing the ore to break or cave under its own weight.

Block model — A model of a mineral deposit that divides it into smaller volumes to allow resource and reserve calculation and mine planning; may be a physical scale model, or a computer construct. **Board lot** – Normally, one hundred shares. Sometimes 500 shares of a penny stock.

Bond – An agreement to pay a certain amount of interest over a given period of time, and to repay the loan on its maturity. Assets are pledged as security.

Boom – A telescoping, hydraulically powered steel arm on which drifters, manbaskets and hydraulic hammers are mounted.

Bottle roll test - A chemical analysis in which a sample is agitated in an extractant solution (cyanide, for example) and the resulting leachate analyzed for its metallic content. Useful in metallurgical testing to predict how much metal can be extracted from an ore.

Box hole – A short raise or opening driven above a drift for the purpose of drawing ore from a stope, or to permit access.

Break – Loosely used to describe a large-scale regional shear zone or structural fault.

Breast – A working face in a mine, usually restricted to a stope.

Breccia – A rock in which angular fragments are surrounded by a mass of finer-grained material. Breccias may form by explosive volcanic action, by structural deformation (a "fault breccia"), by intrusive action (where the intrusive rock incorporates fragments of country rock), or by hydrothermal processes (where wall rock fragments are incorporated by vein material).

Broken reserves – The ore in a mine which has been broken by blasting but which has not yet been transported to surface.

Brunton compass – A pocket compass equipped with sights and a reflector, used for sighting lines, measuring dip and carrying out preliminary surveys.

Bulk mining – Any large-scale, mechanized method of mining involving many thousands of tonnes of ore being brought to surface per day.

Bulk sample – A large sample of mineralized rock, frequently hundreds of tonnes, selected in such a manner as to be representative of the potential orebody being sampled. Used to determine metallurgical characteristics.

Bullion – Metal formed into bars or ingots.

Bull market – Term used to describe financial market conditions when share prices are going up.

Bull quartz – A prospector's term for white, coarse-grained, barren quartz.

Byproduct – A secondary metal or mineral product recovered in the milling process.

C

Cable bolt – A steel cable, capable of withstanding tens of tonnes, cemented into a drillhole to lend support in blocky ground.

Cage – The conveyance used to transport men and equipment between the surface and the mine levels.

Calcine – Name given to concentrate that is ready for smelting (i.e. the sulphur has been driven off by oxidation).

Call – An option to buy shares at a specified price. The opposite of a "put".

Call factor – The fraction of an ore reserve that is successfully mined.

Capitalization – A financial term used to describe the value financial markets put on a company. Determined by multiplying the number of outstanding shares of a company by the current stock price.

Capital stock – The total ownership of a limited liability company divided among a specified number of shares.

Captive stope – A stope that is accessible only through a manway.

Carat – The standard unit of mass used for diamonds and other precious stones, equal to 0.2 gram. (The karat is a unit of concentration.)

Carbon-in-leach (CIL) – A method of recovering gold where activated carbon particles circulate in the leach solution.

Carbon-in-pulp (CIP) – A method of recovering gold and silver from pregnant cyanide solutions by adsorbing the precious metals to granules of activated carbon, which are typically ground up coconut shells.

Cash flow – The net of the inflow and outflow of cash during an accounting period. Does not account for depreciation or bookkeeping write-offs which do not involve an actual cash outlay.

Cathode – A rectangular plate of metal, produced by electrolytic refining, which is melted into commercial shapes such as wirebars, billets, ingots, etc.

Cesium magnetometer – An geophysical instrument which measures magnetic field strength in terms of vertical gradient and total field.

Chalcocite – A sulphide mineral of copper common in the zone of secondary enrichment.

Chalcopyrite – A sulphide mineral of copper and iron; the most common ore mineral of copper.

Change house – The mine building where workers change into work clothes; also known as the "dry".

Channel sample – A sample composed of pieces of vein or mineral deposit that have been cut out of a small trench or channel, usually about 10 cm wide and 2 cm deep.

Chargeability — The effect measured in an induced-polarization survey, defined as the remanent voltage (the voltage that persists after the current is cut off) divided by the applied voltage.

Charter – A document issued by a governing authority creating a company or other corporation.

Chartered bank – A financial institution that accepts deposits and provides loans.

Chip sample – A method of sampling a rock exposure whereby a regular series of small chips of rock is broken off along a line across the face.

Chromite – The chief ore mineral of chromium.

Chute – An opening, usually constructed of timber and equipped with a gate, through which ore is drawn from a stope into mine cars.

Cinnabar – A vermilion-coloured ore mineral of mercury.

Circulating load – Over-sized chunks of ore returned to the head of a closed grinding circuit before going on to the next stage of treatment.

Claim – A portion of land held either by a prospector or a mining company. In Canada, the common size is 1,320 ft. (about 400 m) square, or 40 acres (about 16 ha).

Clarification – Process of clearing dirty water by removing suspended material.

Classifier – A mineral-processing machine which separates minerals according to size and density.

Clay – A fine-grained material composed of hydrous aluminum silicates.

Cleavage – The tendency of a mineral to split along crystallographic planes.

Closed circuit – A loop in the milling process wherein a selected portion of the product of a machine is returned to the head of the machine for finishing to required specification.

Coal – A carbonaceous rock mined for use as a fuel.

Coalification – The metamorphic processes of forming coal.

Collar – The term applied to the timbering or concrete around the mouth of a shaft; also used to describe the top of a drill hole.

Column flotation – A milling process, carried out in a tall cylindrical column, whereby valuable minerals are separated from gangue minerals based on their wetability properties.

Common stock – Shares in a company which have full voting rights which the holders use to control the company in common with each other. There is no fixed or assured dividend as with preferred shares, which have first claim on the distribution of a company's earnings or assets.

Complex ore – An ore containing a number of minerals of economic value. The term often implies that there are metallurgical difficulties in liberating and separating the valuable metals.

Cone crusher – A machine which crushes ore between a gyrating cone or crushing head and an inverted, truncated cone known as a bowl.

Concentrate – A fine, powdery product of the milling process containing a high percentage of valuable metal.

Concentrator – A mill that produces a concentrate of valuable minerals or metals. Further treatment is required to recover the pure metal.

Confirmation – A form delivered by a broker to the client, setting forth the details of stock sales or purchases for the client.

Conglomerate – A sedimentary rock consisting of rounded, water-worn pebbles or boulders cemented into a solid mass.

Contact – A geological term used to describe the line or plane along which two different rock formations meet.

Contact metamorphism – Metamorphism of country rocks adjacent to an intrusion, caused by heat from the intrusion.

Contango – A situation in which the price of a metal for forward or future delivery stands at a premium over the cash or spot price of the metal.

Continuous miner – A piece of mining equipment which produces a continuous flow of ore from the working face.

Controlled blasting – Blasting patterns and sequences designed to achieve a particular objective. Cast blasting, where the muck pile is cast in a particular direction, and deck blasting, where holes are loaded once but blasted in successive blasts days apart, are examples.

Converter – In copper smelting, a furnace used to separate copper metal from matte.

Convertible – A bond or debenture that can be converted (either at the holder's or issuer's option) to shares, according to some predetermined formula and usually on a set schedule.

Core – The long cylindrical piece of rock, about an inch in diameter, brought to surface by diamond drilling.

Core barrel – That part of a string of tools in a diamond drill hole in which the core specimen is collected.

Cordillera – The continuous chain of mountain ranges on the western margin of North and South America.

Country rock – Loosely used to describe the general mass of rock adjacent to an orebody. Also known as the host rock.

Crosscut – A horizontal opening driven from a shaft and (or near) right angles to the strike of a vein or other orebody.

Crown pillar — a body of rock at the top of a mine opening left to support the rock above and to the sides.

Crust – The outermost layer of the Earth; includes both continental and oceanic crust.

Cum-dividend – When the buyer of a share is entitled to a pending dividend payment.

Current assets – Assets of company which can and are likely to be converted into cash within a year. Includes cash, marketable securities, accounts receivable and supplies.

Current liabilities – A company's debts that are payable within one year.

Custom smelter – A smelter which processes concentrates from independent mines. Concentrates may be purchased or the smelter may be contracted to do the processing for the independent company.

Cut-and-fill – A method of stoping in which ore is removed in slices, or lifts, and then the excavation is filled with rock or other waste material (backfill), before the subsequent slice is extracted.

Cut value – Applies to assays that have been reduced to some arbitrary maximum to prevent erratic high values from inflating the average.

Cutoff grade - A grade below which samples are not included in a resource or reserve.

Cyanidation – A method of extracting exposed gold or silver grains from crushed or ground ore by dissolving it in a weak cyanide solution. May be carried out in tanks inside a mill or outside in heaps.

Cyanide – A chemical species containing carbon and nitrogen used to dissolve gold and silver from ore.

D

Day order – An order to buy or sell shares, good only on the day the order was entered.

Debenture – An indebtedness backed only by the general credit of the issuer and unsecured by a lien on any specific asset.

Debt financing – Method of raising capital whereby companies borrow money from a lending institution.

Deck – The area around the shaft collar where men and materials enter the cage to be lowered underground.

Decline – A sloping underground opening for machine access from level to level or from surface; also called a ramp.

Deferred charges – Expenses incurred but not charged against the current year's operation.

Density – The mass of a substance per unit volume. In resource or reserve estimation, it is expressed in tonnes per cubic metre or tons per cubic foot; when working with smaller substances, it is more often expressed in grams per cubic centimetre. Also called specific gravity.

Depletion – An accounting device, used primarily in tax computations. It recognizes the consumption of an ore deposit, a mine's principal asset.

Deposit – A body of rock containing valuable minerals; usage generally restricted to zones of mineralization whose size has been wholly or partly determined through sampling.

Depreciation – The periodic, systematic charging to expense of plant assets reflecting the decline in economic potential of the assets.

Development – Underground work carried out for the purpose of opening up a mineral deposit. Includes shaft sinking, crosscutting, drifting and raising.

Development drilling – drilling to establish accurate estimates of mineral reserves.

Diabase – A common basic igneous rock usually occurring in dykes or sills.

Diamond – The hardest known mineral, composed of pure carbon; low-quality diamonds are used to make bits for diamond drilling in rock.

Diamond drill – A rotary type of rock drill that cuts a core of rock that is recovered in long cylindrical sections, two cm or more in diameter.

Diamond driller – A person who operates a diamond drill.

Dilution (mining) – The necessary mining of waste rock along with ore in underground mining; the ratio of waste rock to the total amount of rock mined in a stope or mine, usually expressed as a percentage.

Dilution (of shares) – Diminution of share value through issuing shares or options to buy shares, which decreases the actual or imputed earnings per share.

Diorite – An intrusive igneous rock composed chiefly of sodic plagioclase, hornblende, biotite or pyroxene.

Dip – The angle at which a vein, structure or rock bed is inclined from the horizontal as measured at right angles to the strike.

Dip needle – A compass with the needle mounted so as to swing in a vertical plane, used for prospecting to determine the magnetic attraction of rocks.

Directional drilling – A method of drilling involving the use of stabilizers and wedges to direct the orientation of the hole.

Discounted cash flow – A technique for estimating the value and profitability of a project, based on the assumed revenue from production, capital and operating costs, and the potential return from a risk-free investment (the "discount rate").

Disseminated ore – Ore carrying small particles of valuable minerals spread more or less uniformly through the host rock.

Dividend – Cash or stock awarded to preferred and common shareholders at the discretion of the company's board of directors.

Dividend claim – Made when a dividend has been paid to the previous holder because stock has not yet been transferred to the name of the new owner.

Doré bar – The final saleable product of a gold mine. Usually consisting of gold and silver.

Drag fold – The result of the plastic deformation of a rock unit where it has been folded or bent back on itself.

Drawpoint – An underground opening at the bottom of a stope through which broken ore from the stope is extracted.

Drift – A horizontal underground opening that follows along the length of a vein or rock formation as opposed to a crosscut which crosses the rock formation.

Drift-and-fill — a cut-and-fill mining method taking only half the width of the stope at one time.

Drifter – A hydraulic rock drill used to drill small-diameter holes for blasting or for installing rock bolts.

Drill-indicated resources – The size and quality of a potential orebody as suggested by widely spaced drillholes; more work is required before resources can be classified as probable or proven reserves.

Dry – A building where the miner changes into working clothes.

Due diligence – The care and caution required before making a decision; loosely, a financial and technical investigation to determine whether an investment is sound.

Dump – A pile of broken waste rock or ore on surface.

Dyke – A long and relatively thin body of igneous rock that, while in the molten state, intruded a fissure in older rocks.

E

Electrolysis – An electric current is passed through a solution containing dissolved metals, causing the metals to be deposited onto a cathode.

Electrolytic refining – The process of purifying metal ingots that are suspended as anodes in an electrolytic bath, alternated with refined sheets of the same metal which act as starters or cathodes.

EM survey – A geophysical survey method which measures the electromagnetic properties of rocks. **Emulsion explosive** — an explosive mixture containing a water solution of an oxidant plus a fuel (petroleum, metal powder, or an organic compound), plus an emulsifier to keep the constituents from separating.

En echelon – Roughly parallel but staggered structures.

Environmental impact study – A written report, compiled prior to a production decision, that examines the effects proposed mining activities will have on the natural surroundings.

Epigenetic – Orebodies formed by hydrothermal fluids and gases that were introduced into the host rocks from elsewhere, filling cavities in the host rock.

Epithermal deposit – A mineral deposit consisting of veins and replacement bodies, usually in volcanic or sedimentary rocks, containing precious metals or, more rarely, base metals.

Equity financing – The acquisition of funds by selling treasury shares.

Era – A large division of geologic time - the Precambrian era, for example (see Appendix IV).

Erosion – The breaking down and subsequent removal of either rock or surface material by wind, rain, wave action, freezing and thawing and other processes.

Erratic – Either a piece of visible gold or a large glacial boulder.

Escrowed shares – Shares deposited in trust pending fulfilment of certain conditions, and not ordinarily available for trading until released.

Ex-dividend – On stocks selling "ex-dividend", the seller retains the right to a pending dividend payment.

Exploration – Prospecting, sampling, mapping, diamond drilling and other work involved in searching for ore.

F

Face – The end of a drift, crosscut or stope in which work is taking place.

Fault – A break in the Earth's crust caused by tectonic forces which have moved the rock on one side with respect to the other.

Feasibility study – An economic study assessing whether a mineral deposit can be mined profitably, by estimating the capital and operating costs of a mine and the potential revenues from production.

Feldspar – A group of common rock-forming minerals that includes microcline, orthoclase, plagioclase and others.

Felsic – Term used to describe light-colored rocks containing feldspar, feldspathoids and silica.

Ferrous – Containing iron.

Fineness – The proportion of pure gold or silver in jewelry or bullion expressed in parts per thousand. Thus, 925 fine gold indicates 925 parts out of 1,000, or 92.5% is pure gold.

Fissure – An extensive crack, break or fracture in rocks.

Fixed Assets – Possessions such as buildings, machinery and land which, as opposed to current assets, are unlikely to be converted into cash during the normal business cycle.

Float – Pieces of rock that have been broken off and moved from their original location by natural forces such as frost or glacial action.

Flotation – A milling process in which valuable mineral particles are induced to become attached to bubbles and float as others sink.

Flowsheet – An illustration showing the sequence of operations, step by step, by which ore is treated in a milling, concentration or smelting process.

Flow-through shares – Shares in an exploration company that allow the tax deduction or credits for mineral exploration to be passed from the company to the shareholder.

Flux – A chemical substance that reacts with gangue minerals to form slags, which are liquid at furnace temperature and low enough in density to float on the molten bath of metal or matte.

Fluxgate magnetometer – An instrument used in geophysics to measure total magnetic field.

Fold – Any bending or wrinkling of rock strata.

Footwall – The rock on the underside of a vein or ore structure.

Forward contract – The sale or purchase of a commodity for delivery at a specified future date.

Fracture – A break in the rock, the opening of which allows mineral-bearing solutions to enter. A "cross-fracture" is a minor break extending at more-or-less right angles to the direction of the principal fractures.

Fragmental rock – A rock made up of fragments or clasts, rather than of mineral grains that have crystallized together. More narrowly, a volcanic rock made up of fragments, such as a tuff or agglomerate. In that latter usage, it can also be called a pyroclastic rock.

Free milling – Ores of gold or silver from which the precious metals can be recovered by concentrating methods without resorting to pressure leaching or other chemical treatment.

Fully diluted (of shares or earnings) – Based on the maximum number of shares a company could have, including all shares that would be issued if convertible debt, rights, warrants, and options were exercised.

G

GAAP – Generally Accepted Accounting Principles; the conventions by which financial accounting is done. Two widely applied GAAP systems are those of the International Accounting Standards Board, prominent in Europe, and the U.S. Financial Accounting Standards Board.

Gabbro – A dark, coarse-grained igneous rock.

Galena – Lead sulphide, the most common ore mineral of lead.

Gamma – A unit of measurement of magnetic intensity.

Gangue – The worthless minerals in an ore deposit.

Geiger counter – An instrument used to measure the radioactivity that emanates from certain minerals by means of a Geiger-Mueller tube.

Geochemistry – The study of the chemical properties of rocks.

Geology – The science concerned with the study of the rocks which compose the Earth.

Geophysics – The study of the physical properties of rocks and minerals.

Geophysical survey – A scientific method of prospecting that measures the physical properties of rock formations. Common properties investigated include magnetism, specific gravity, electrical conductivity and radioactivity.

Geostatistics — A technique used to estimate the grades and tonnages of mineral deposits through assuming a correlation of grades with position, and interpolating between grades at different sample locations.

Geothermal – Pertains to the heat of the Earth's interior.

Glacial drift – Sedimentary material that has been transported by glaciers.

Glacial striations – Lines or scratches on a smooth rock surface caused by glacial abrasion.

Glory hole – An open pit from which ore is extracted, especially where broken ore is passed to underground workings before being hoisted.

Gneiss – A layered or banded crystalline metamorphic rock, the grains of which are aligned or elongated into a roughly parallel arrangement.

Gold loan – A form of debt financing whereby a potential gold producer borrows gold from a lending institution, sells the gold on the open market, uses the cash for mine development, then pays back the gold from actual mine production.

Gossan – The rust-colored capping or staining of a mineral deposit, generally formed by the oxidation or alteration of iron sulphides.

Gouge – Fine, putty-like material composed of ground-up rock found along a fault.

Grab sample – A sample from a rock outcrop that is assayed to determine if valuable elements are contained in the rock. A grab sample is not intended to be representative of the deposit, and usually the best-looking material is selected.

Graben – A downfaulted block of rock.

Grade – The concentration of metal or valuable mineral in a body of rock, usually expressed as a percentage or in grams per tonne or ounces per ton; also, the concentration of metal in a mill concentrate or matte.

Granite – A coarse-grained intrusive igneous rock consisting of quartz, feldspar and mica.

Gravity meter, gravimeter – An instrument for measuring the gravitational attraction of the earth; gravitational attraction varies with the density of the rocks in the vicinity.

Greenstone belt – An area underlain by metamorphosed volcanic and sedimentary rocks, usually in a continental shield.

Grey market – The market for securities that are not yet listed on an exchange or traded in a quotation system.

Grizzly (or mantle) – A grating, usually constructed of steel rails, placed over the top of a chute or ore pass for the purpose of stopping large pieces of rock or ore that may hang up in the pass.

Gross value – The theoretical value of ore determined simply by applying the assay of metal or metals and the current market price. It must be used only with caution and severe qualification.

Gross value royalty – A share of gross revenue from the sale of minerals from a mine.

Grouting – The process of sealing off a water flow in rocks by forcing a thin slurry of cement or other chemicals into the crevices; usually done through a diamond drill hole.

Grubstake – Finances or supplies of food, etc., furnished to a prospector in return for an interest in any discoveries made.

Guides – The timber rails installed along the walls of a shaft for steadying, or guiding, the cage or conveyance.

Gypsum – A sedimentary rock consisting of hydrated calcium sulphate.

Gyratory crusher – A machine that crushes ore between an eccentrically mounted crushing cone and a fixed crushing throat. Typically has a higher capacity than a jaw crusher.

H

Halite – Rock salt.

Hangingwall – The rock on the upper side of a vein or ore deposit.

Headframe - The structure built above a mine shaft for the hoisting system.

Head grade – The average grade of ore fed into a mill.

Heap leaching – A process whereby valuable metals, usually gold and silver, are leached from a heap, or pad, of crushed ore by leaching solutions percolating down through the heap and collected from a sloping, impermeable liner below the pad.

Hedging – Taking a buy or sell position in a futures market opposite to a position held in the cash market to minimize the risk of financial loss from an adverse price change.

Hematite – An oxide of iron, and one of that metal's most common ore minerals.

High grade – Rich ore. As a verb, it refers to selective mining of the best ore in a deposit.

High-grader – One who mines only the richest ore, especially gold.

Hoist – The machine used for raising and lowering the cage or other conveyance in a shaft.

Holding company – A corporation engaged principally in holding a controlling interest in one or more other companies.

Hornfels – A fine-grained contact metamorphic rock.

Horse – A mass of waste rock lying within a vein or orebody.

Horst – An upfaulted block of rock.

Host rock – The rock surrounding an ore deposit.

Hydrometallurgy – The treatment of ore by wet processes, such as leaching, resulting in the solution of a metal and its subsequent recovery.

Hydrothermal – Relating to hot fluids circulating in the earth's crust.

Hypogene - Processes occurring at depth; especially, the primary hydrothermal processes that formed a mineral deposit.

I

Igneous rocks – Rocks formed by the solidification of molten material from far below the earth's surface.

Ilmenite – An ore mineral of titanium, being an iron-titanium oxide.

Indicated resource – A resource whose size and grade have been estimated from sampling at places spaced closely enough that its continuity can be reasonably assumed.

Indicator minerals — Minerals typical of particular rock types or mineral deposits, for example, pyrope garnets and chromium diopsides from kimberlites. Tracing indicator minerals back to their source can be a valuable method for finding mineral deposits.

Induced polarization – A method of ground geophysical surveying employing an electrical current to determine indications of mineralization.

Inductively coupled plasma (ICP) — A loose term for a type of atomic-emission spectrometry, an instrumental method of chemical analysis that detects metallic atoms by the wavelengths of light they emit when energized. (ICP properly refers only to the excitation source and not to the analytical method itself). An efficient and widely used technique for assaying and geochemical analysis of metals.

Industrial minerals – Non-metallic, non-fuel minerals used in the chemical and manufacturing industries. Examples are asbestos, gypsum, salt, graphite, mica, gravel, building stone and talc.

Inferred resource – A resource whose size and grade have been estimated mainly or wholly from limited sampling data, assuming that the mineralized body is continuous based on geological evidence.

Initial public offering – The first sale of shares to the public, usually by subscription from a group of investment dealers.

Installment receipt – A security representing newly issued stock, carrying the obligation that further payments are required before the actual stock will be issued.

Institutional investors – Pension funds and mutual funds, managing money for a large number of individual investors.

Intermediate rock – An igneous rock containing 52% to 66% quartz.

Intrusion – A body of igneous rock formed by the consolidation of magma intruded into other rocks, in contrast to lavas, which are extruded upon the surface.

Inverse distance weighting — A mathematical procedure for interpolation of grades between sample locations; grade values at nearby sample locations are weighted heavily, those at distant sample locations weighted negligibly, and a weighted average grade calculated.

Ion exchange – An exchange of ions in a crystal with irons in a solution. Used as a method for recovering valuable metals, such as uranium, from solution.

J

Jaw crusher – A machine in which rock is broken by the action of steel plates.

Jig – A piece of milling equipment used to concentrate ore on a screen submerged in water, either by the reciprocating motion of the screen or by the pulsation of water through it.

Joint Ore Reserves Committee code (JORC code) — A code developed by the Australian mineral industry for consistent reporting of resource and reserve estimates.

K

Karat – A unit of concentration, equal to one part in 24, sometimes used instead of fineness to express the purity of precious metals.

Kimberlite – A variety of peridotite; the most common host rock of diamonds.

Kriging — In geostatistics, estimation of the grade at an unsampled location from known or estimated grades at other locations nearby.

L

Lagging – Planks or small timbers placed between steel ribs along the roof of a stope or drift to prevent rocks from falling, rather than to support the main weight of the overlying rocks.

Lamprophyre – An igneous rock, composed of dark minerals, that occurs in dykes; sometimes contains diamonds.

Laterite – A residual soil, ususally found in tropical countries, out of which the silica has been leached. May form orebodies of iron, nickel, cobalt, bauxite and manganese.

Launder – A chute or trough for conveying pulp, water or powdered ore in a mill.

Lava – A general name for the molten rock ejected by volcanoes.

Leachable – Extractable by chemical solvents.

Leaching – A chemical process for the extraction of valuable minerals from ore; also, a natural process by which ground waters dissolve minerals, thus leaving the rock with a smaller proportion of some of the minerals than it contained originally.

Lens – Generally used to describe a body of ore that is thick in the middle and tapers towards the ends.

Lenticular – A deposit having roughly the form of a double convex lens.

Level – The horizontal openings on a working horizon in a mine; it is customary to work mines from a shaft, establishing levels at regular intervals, generally about 50 metres or more apart.

Lignite – A soft, low-rank, brownish-black coal.

Limestone – A bedded, sedimentary deposit consisting chiefly of calcium carbonate.

Limit order – An order made by a client to a broker to buy or sell shares at a specified price or better.

Limonite – A brown, hydrous iron oxide.

Line cutting – Straight clearings through the bush or jungle to permit sightings for geophysical and other surveys.

Loading pocket – A chamber excavated in the rock at the base of an ore pass, where broken ore is stored in preparation for hoisting to the surface.

Lode – A mineral deposit in solid rock.

Logging – The process of recording geological observations of drill core either on paper or on computer disk.

London fix – The twice-daily bidding session held by five dealing companies to set the gold price. There are also daily London fixes to set the prices of other precious metals.

London Metals Exchange – A major bidding market for base metals, which operates daily in London.

Long position – Securities owned outright or carried on margin.

Long ton – 2,240 lbs. avoirdupois (compared with a short ton, which is 2,000 lbs.).

Longwall – A mining method where narrow vertical slices of ore are cut along long straight faces or walls.

M

Macrodiamond — A diamond grain, usually defined as having at least one dimension (length, width, or height) greater than 0.5 mm (compare microdiamond).

Mafic – Igneous rocks composed mostly of dark, iron- and magnesium-rich minerals.

Magma – The molten material deep in the Earth from which igneous rocks are formed.

Magmatic segregation – An ore-forming process whereby valuable minerals are concentrated by settling out of a cooling magma.

Magnetic gradient survey – A geophysical survey using a pair of magnetometers a fixed distance apart, to measure the difference in the magnetic field with height above the ground.

Magnetic separation – A process in which a magnetically susceptible mineral is separated from gangue minerals by applying a strong magnetic field; ores of iron are commonly treated in this way.

Magnetic susceptibility – A measure of the degree to which a rock is attracted to a magnet.

Magnetic survey – A geophysical survey that measures the intensity of the Earth's magnetic field.

Magnetite – Black, magnetic iron ore, an iron oxide.

Magnetometer – An instrument used to measure the magnetic attraction of underlying rocks.

Map-staking – A form of claim-staking practised in some jurisdictions whereby claims are staked by drawing lines around the claim on claim maps at a government office.

Marble – A metamorphic rock derived from the recrystallization of limestone under intense heat and pressure.

Margin – Cash deposited with a broker as partial payment of the purchase price for any type of listed stock. The stock is held by the broker as security for the loan.

Marginal deposit – An orebody of minimal profitability.

Market order – An order to buy or sell at the best price available. In absence of any specified price or limit, an order is considered to be "at the market".

Mass Spectrometry (MS) — An instrumental method of chemical analysis that separates and detects atoms by their mass. Can be used to measure very low concentrations of an element in a sample, and also to determine the ratio of isotopes of a single element in a sample.

Massive sulphide – A body of rock made up mainly or wholly of sulphide minerals, such as pyrite, pyrrhotite, or chalcopyrite; often proves to be an orebody. Also, a mineral deposit occurring in massive-sulphide form.

Matte – A product of a smelter, containing metal and some sulphur, which must be refined further to obtain pure metal.

Measured resource – A resource whose size and grade have been estimated from sampling at places spaced closely enough that its continuity is essentially confirmed.

Mesh size — A system of expressing particle sizes based on the standard openings of wire mesh sieves; for example, 200 mesh is equal to 0.074 mm. As a crude rule of thumb, dividing 15 by the mesh number gives an approximate particle size in mm.

Metallurgical coal – Coal used to make steel.

Metallurgy – The study of extracting metals from their ores.

Metamorphic rocks – Rocks which have undergone a change in texture or composition as the result of heat and/or pressure.

Metamorphism – The process by which the form or structure of rocks is changed by heat and/or pressure.

Metasomatism — Replacement of pre-existing minerals and mineral aggregates in a rock by newly introduced material, usually transported in water solutions moving through fractures and porous zones in the rock; the chemical reactions between the rock and the solutions create new minerals. One of the principal agencies of chemical and mineralogical alteration; sometimes an ore-forming process itself.

Microdiamond — A small diamond grain, usually defined as being less than 0.5 mm in its longest dimension. Often found in exploration work or bulk sampling, and indicates that a rock hosts diamonds; not economically extractable itself (compare macrodiamond).

Micron (µm) — Now correctly called a micrometre; one one-millionth of a metre, or 0.001 mm. Often used to express grain size.

Microprobe — An analytical instrument that can determine the composition of an individual mineral grain, or even of parts of a grain, by measuring the energy emitted by the grain when hit by a beam of electrons. Frequently used in diamond exploration to determine the metal ratios in diamond indicator minerals.

Migmatite – Rock consisting of thin, alternating layers of granite and schist.

Mill – A plant in which ore is treated and metals are recovered or prepared for smelting; also a revolving drum used for the grinding of ores in preparation for treatment.

Milling ore – Ore that contains sufficient valuable mineral to be treated by milling process.

Millivolts – A measure of the voltage of an electric current, specifically, one-thousandth of a volt.

Minable reserves – A redundant term, since reserves, by definition, comprise only the portion of a deposit that is minable at a profit.

Mineral – A naturally occurring homogeneous substance having definite physical properties and chemical composition and, if formed under favorable conditions, a definite crystal form.

Minority interest – That part of a company's equity held by shareholders that do not hold the controlling interest in the company; in subsidiary companies, the fraction of assets or earnings that accrue to the subsidiary's minority shareholders.

Mobile metal ion geochemistry (MMI) — A geochemical technique in which weakly bound metals, transported upward from deeper mineral deposits, are extracted from a soil and analyzed.

Muck – Ore or rock that has been broken by blasting.

Muck sample – A representative piece of ore that is taken from a muck pile and then assayed to determine the grade of the pile.

Multispectral scanning — A remote-sensing technique, used mainly from satellite or aircraft platforms, which maps the reflection of multiple wavelengths of visible light and infrared radiation from the earth's surface. The images produced by the scanner can be used to map alteration and structural patterns. Also called "hyperspectral scanning" if it records hundreds of wavelength channels.

N

Nanotesla – The international unit for measuring magnetic flux density.

National Instrument — In Canadian securities law, a regulation approved by securities regulators in all provinces and adopted into law under provincial Securities Acts. For example, National Instrument 43-101 governs disclosure of information from mineral exploration and development; Instrument 51-101 serves the same function in the oil and gas industry.

Native metal – A metal occurring in nature in pure form, uncombined with other elements.

Net asset value – A corporation's total assets, minus its total liabilities; also called "shareholder's equity" or "book value."

Net present value – A value calculated for a security or a project, based on its assumed future revenue and discounted by the risk-free return.

Net profit interest – A portion of the profit remaining after all charges, including taxes and bookkeeping charges, such as depreciation, have been deducted.

Net smelter return – The gross value of metal in an ore or concentrate, minus transportation, smelting and refining charges; a royalty based on a net smelter return value.

Net worth – The difference between total assets and total liabilities.

Neutron activation analysis — An instrumental method of chemical analysis in which the sample is irradiated with neutrons, producing radioisotopes of the elements being analyzed. The radiation emitted by the isotopes is measured to determine the concentration of elements in the sample. A highly accurate and sensitive analytical method.

No par value – A phrase indicating a stock has no specified value at the time of its issue.

Norite – A coarse-grained igneous rock that is host to copper/nickel deposits in the Sudbury area of Ontario.

Nugget – A large mass of precious metal, found free in nature.

Nugget effect — The tendency of small samples to understate the concentration of metals that occur in discrete particles (for example, gold or diamonds). Can be surmounted by taking larger samples.

O

Odd lot – A block of shares that is less than a board lot.

Open order – An order to buy or sell stock, which is good until cancelled by the client.

Open pit – A mine that is entirely on surface. Also referred to as open-cut or open-cast mine.

Open stoping – Mining by removing a relatively large block of ore and allowing the void to remain open while mining is completed; an open stope may later be backfilled.

Option – An agreement to purchase a property reached between the property vendor and some other party who wishes to explore the property further.

Option (on stock) – The right to buy or sell a share at a set price, regardless of market value.

Ore – A mixture of ore minerals and gangue from which at least one of the ore minerals can be extracted at a profit.

Ore pass – Vertical or inclined passage for the downward transfer of ore connecting a level with the hoisting shaft or a lower level.

Orebody – A natural concentration of a valuable mineral or minerals that can be extracted and sold at a profit.

Ore Reserves – The calculated tonnage and grade of mineralization which can be extracted profitably; classified as probable and proven according to the level of confidence that can be placed in the data.

Oreshoot – The portion, or length, of a vein or other structure that carries sufficient valuable minerals to be extracted profitably.

Organic maturation – The process of turning peat into coal.

Orogeny – A period of mountain-building characterized by the folding of a portion of the earth's crust.

Orthogneiss — A gneiss with a high content of quartz, feldspars, and mafic minerals, suggesting it was derived from an igneous parent rock.

Outcrop – An exposure of rock or mineral deposit that can be seen on surface, that is, not covered by soil or water.

Over-the-counter market – The market in securities that are not traded on stock exchanges or through quotation systems.

Overturned – Where the oldest sedimentary rock beds are lying on top of a younger beds.

Oxidation – A chemical reaction caused by exposure to oxygen that results in a change in the chemical composition of a mineral.

Oxide – A compound of oxygen and some other element; iron, titanium, manganese and tin ores are all oxides.

Oxide ore – Ore or mineralization in the zone of weathering, where sulphide minerals have broken down to oxides through exposure to the atmosphere and groundwater (see "sulphide ore").

P

Pan – To wash gravel, sand or crushed rock samples in order to isolate gold or other valuable metals by their higher density.

Paragneiss — A gneiss with a high content of alumino-silicate miner-als, such as sillimanite, cordierite, or micas, presumed to be derived from a sedimentary parent rock.

Participating interest – A company's interest in a mine, which entitles it to a certain percentage of profits in return for putting up an equal percentage of the capital cost of the project.

Par value – The stated face value of a stock. Par value shares have no specified face value, but the total amount of authorized capital is set down in the company's charter.

Passing size — In mineral processing or metallurgical testing, the upper size limit of a specified fraction of the particles in a process. For example, a grinding mill might produce a mixture of particles where 80% were smaller than 0.1 mm; this would be expressed as a passing size of P_{80}=0.1 mm.

Patent – The ultimate stage of holding a mineral claim, after which no more assessment work is necessary because all mineral rights have been earned.

Pegmatite – A coarse-grained, igneous rock, generally coarse, but irregular in texture, and similar to a granite in composition; usually occurs in dykes or veins and sometimes contains valuable minerals.

Pellet – A marble-sized ball of iron ore fused with clay for transportation and use in steelmaking.

Pentlandite – Nickel iron sulphide, the most common nickel ore.

Peridotite – An intrusive igneous rock consisting mainly of olivine.

Permeability — A rock's capacity to transmit fluids, such as groundwater. The "coefficient of permeability," now more correctly called the hydraulic conductivity, is the volume of water a rock will allow to flow in unit time across a unit area, most often measured in cubic metres per square metre per second.

Phaneritic – A term used to describe the coarse-grained texture of some igneous rocks.

Picket line – A reference line, marked by pickets or stakes, established on a property for mapping and survey purposes.

Pig iron – Crude iron from a blast furnace.

Pillar – A block of solid ore or other rock left in place to structurally support the shaft, walls or roof of a mine.

Pilot plant - A small-scale metallurgical plant built to test a process.

PIMA — Portable infrared multispectral analyzer; an instrument measuring infrared radiation from the surfaces of rock samples or outcrops. Particular minerals reflect specific wavelengths of infrared radiation.

Pitch - An obsolete term formerly used as a synonym either for plunge or for rake.

Pitchblende – An important uranium ore mineral. It is black in color, possesses a characteristic greasy lustre and is highly radioactive.

Placer – A deposit of sand and gravel containing valuable metals such as gold, tin or diamonds.

Plant – A building or group of buildings in which a process or function is carried out; at a mine site it will include warehouses, hoisting equipment, compressors, maintenance shops, offices and the mill or concentrator.

Plate tectonics – A geological theory which postulates that the Earth's crust is made up of a number of rigid plates which collide, rub up against and spread out from one another.

Plug – A common name for a small offshoot from a large body of molten rock.

Plunge – The vertical angle a linear geological feature makes with the horizontal plane.

Plutonic – Refers to rocks of igneous origin that have come from great depth.

Point – Unit of value of a stock as quoted by a stock exchange. May represent one dollar, one cent or one-eighth of a dollar, depending on the stock exchange.

Poison pill - A provision in a company's by-laws allowing the board to issue additional stock to existing shareholders at a price below the market price; used to dissuade the bidder in a takeover battle.

Polishing pond – The last in a series of settling ponds through which mill effluent flows before being discharged into the natural environment.

Polygonal estimate — A resource or reserve estimate calculated from the volume, average grade, and average density of a series of polygonal blocks, whose size and shape approximates the form of the mineral deposit.

Pooling shares – See escrowed shares.

Porosity — The ratio of pore volume to the total volume of a rock, usually expressed as a percentage (compare permeability).

Porphyry – Any igneous rock in which relatively large crystals, called phenocrysts, are set in a fine-grained groundmass.

Porphyry copper – A deposit of disseminated copper minerals in or around a large body of intrusive rock.

Portal – The surface entrance to a tunnel or adit.

Portfolio – A collection of financial assets.

Potash – Potassium compounds mined for fertilizer and for use in the chemical industry.

Precambrian shield – The oldest, most stable regions of the earth's crust, the largest of which is the Canadian Shield.

Preferred shares – Shares of a limited liability company that rank ahead of common shares, but after bonds, in distribution of earnings or in claim to the company's assets in the event of liquidation. They pay a fixed dividend but normally do not have voting rights, as with common shares.

Price-to-earnings ratio – The current market price of a stock divided by the company's net earnings per share for the year.

Primary deposits – Valuable minerals deposited during the original period or periods of mineralization, as opposed to those deposited as a result of alteration or weathering.

Private placement – Sale of shares to individuals or corporations outside the normal market, at a negotiated price. Often used to raise capital for a junior exploration company.

Pro rata – In proportion, usually to ownership, income or contribution.

Probable reserves – Valuable mineralization not sampled enough to be termed "proven."

Profit and loss statement – The income statement of a company detailing revenues minus total costs to give total profit.

Prospect – A mineral occurrence that is being, or has been, explored; often restricted to mineral occurrences that have been drilled.

Prospectus – A document filed with the appropriate securities commission detailing the activities and financial condition of a company seeking funds from the public through the issuance of shares.

Proton precession magnetometer – A geophysical instrument which measures magnetic field intensity in terms of vertical gradient and total field.

Proven reserves – Reserves that have been sampled extensively by closely spaced diamond drill holes and possibly developed by underground workings in sufficient detail to render an accurate estimation of grade and tonnage.

Proxy – A power of attorney given by the shareholder so that his or her stock may be voted by his/her nominee(s) at shareholders' meetings.

Pulp – Pulverized or ground ore in solution.

Put – An option to sell a stock at an agreed upon price within a specified time. The owner can present his put to the contracting broker at any time within the option period and compel him to buy the stock.

Pyramiding – The use of increased buying power to increase ownership arising from price appreciation.

Pyrite – A yellow iron sulphide mineral, normally of little value. It is referred to as "fool's gold".

Pyroclastic rock – A volcanic fragmental rock, such as a tuff, lapilli tuff, agglomerate or breccia.

Pyrrhotite – A bronze-colored, magnetic iron sulphide mineral.

Q

Quartz – Common rock-forming mineral consisting of silicon and oxygen.

Quartzite – A metamorphic rock formed by the transformation of a sandstone by heat and pressure.

R

Radioactivity – The property of spontaneously emitting alpha, beta or gamma rays by the decay of the nuclei of atoms.

Radon survey – A geochemical survey technique which detects traces of radon gas, a product of radioactivity.

Raise – A vertical or inclined underground working that has been excavated from the bottom upward.

Raise borer – A machine that uses a rotating boring head to open a raise.

Rake – The plunge of an orebody measured in the plane of the structure it lies in.

Rare earth elements – Thirty elements composed of the lanthanide and actinide series.

Reaming shell – A component of a string of rods used in diamond drilling, it is set with diamonds and placed between the bit and the core barrel to maintain the gauge (or diameter) of the hole.

Reclamation – The restoration of a site after mining or exploration activity is completed.

Reconnaissance – A preliminary survey of ground.

Record date – The date by which a shareholder must be registered on the books of a company in order to receive a declared dividend, or to vote on company affairs.

Recovery – The percentage of valuable metal in the ore that is recovered by metallurgical treatment.

Refractory ore – Ore that resists the action of chemical reagents in the normal treatment processes and which may require pressure leaching or other means to effect the full recovery of the valuable minerals.

Regional metamorphism – Metamorphism caused by both the heat of igneous processes and tectonic pressure.

Regolith — Loose rock material overlying bedrock, either weathered in place or transported by surface processes.

Remote sensing — The use of photographic and other imagery taken from aircraft or satellites to map geological structures and other characteristics of the earth's surface.

Replacement ore – Ore formed by a process during which certain minerals have passed into solution and have been carried away, while valuable minerals from the solution have been deposited in the place of those removed.

Reserve – that part of a mineral resource that can be mined profitably.

Resistivity survey – A geophysical technique used to measure the resistance of a rock formation to an electric current.

Resource – The calculated amount of material in a mineral deposit, classified as measured, indicated, or inferred, based on the density of drill hole information used.

Resuing – A method of stoping in narrow-vein deposits whereby the wallrock on one side of the vein is blasted first and then the ore.

Reverberatory furnace – A long, flat furnace used to slag gangue minerals and produce a matte.

Reverse circulation – A drilling method in which a rotating bit cuts rock or compacted earth into fragments, which are flushed upward to the drill collar by water or fluid mixtures for sampling. Unlike diamond drilling, it does not provide an intact core for examination or sampling.

Rhyolite – A fine-grained, extrusive igneous rock which has the same chemical composition as granite.

Rib samples – Ore taken from rib pillars in a mine to determine metal content.

Rights – In finance, a certified right to purchase treasury shares in stated quantities, prices and time limits; usually negotiable at a price which is related to the prices of the issue represented, but with a shorter term than a warrant. Rights and warrants can be bought and sold prior to their expiry date because not all shareholders wish to exercise their rights.

Rock – Any natural combination of minerals; part of the earth's crust.

Rockbolting – The act of supporting openings in rock with steel bolts anchored in holes drilled especially for this purpose.

Rockburst – A violent release of energy resulting in the sudden failure of walls or pillars in a mine, caused by the weight or pressure of the surrounding rocks.

Rock factor – The number of cubic metres of a particular rock type required to make up one tonne of the material. One tonne of a highly lateritic ore may occupy 1.0 cubic metre, while a tonne of dense sulphide ore may occupy only 0.25 cubic metre. The reciprocal of density or specific gravity.

Rock mechanics – The study of the mechanical properties of rocks, which includes stress conditions around mine openings and the ability of rocks and underground structures to withstand these stresses.

Rod mill – A rotating steel cylinder that uses steel rods as a means of grinding ore.

Room-and-pillar mining – A method of mining flat-lying ore deposits in which the mined-out areas, or rooms, are separated by pillars of approximately the same size.

Rotary air blast – A rotary drilling method that uses compressed air to move drill cuttings up to the drill collar.

Rotary drill – A machine that drills holes by rotating a rigid, tubular string of drill rods to which is attached a bit. Commonly used for drilling large-diameter blastholes in open-pit mines.

Royalty – An amount of money paid at regular intervals by the lessee or operator of a mining property to a lender or the owner of the ground. Generally based on a certain amount per tonne or a percentage of the total production or profits. Also, the fee paid for the right to use a patented process.

Run-of-mine – A term used loosely to describe ore of average grade.

S

Salting – Fraudulently introducing metals or minerals into a deposit or samples, resulting in falsely higher assays.

Sample – A small portion of rock or a mineral deposit taken so that the metal content can be determined by assaying.

Sampling – Selecting a fractional but representative part of a mineral deposit for analysis.

Sandstone – A sedimentary rock consisting of grains of sand cemented together.

Saprolite - A strongly weathered residual soil formed in place from rock, and made up mainly of clay and silt, but unlike laterite, retaining its silica component. Further weathering may turn it into a laterite, and it usually occurs as a layer between laterite and the unweathered bedrock.

Scaling – The act of removing loose slabs of rock from the back and walls of an underground opening, usually done with a hand-held scaling bar or with a boom-mounted scaling hammer.

Scarp – An escarpment, cliff or steep slope along the margin of a plateau, mesa or terrace.

Schist – A foliated metamorphic rock the grains of which have a roughly parallel arrangement; generally developed by shearing.

Scintillation counter – An instrument used to detect and measure radioactivity by detecting gamma rays; more sensitive than a geiger counter.

Scoping study — An early-stage study on the economics of a mining project used for development planning. It is generally based on assumptions and estimated costs, and is neither as detailed nor as reliable as a feasibility study (for example, it cannot be presented to a lender). May also be called a "preliminary economic assessment".

Secondary enrichment – Enrichment of a vein or mineral deposit by minerals that have been taken into solution from one part of the vein or adjacent rocks and redeposited in another.

Sedimentary rocks – Secondary rocks formed from material derived from other rocks and laid down under water. Examples are limestone, shale and sandstone.

Seismic prospecting – A geophysical method of prospecting, utilizing knowledge of the speed of reflected sound waves in rock.

Self-potential – A technique, used in geophysical prospecting, which recognizes and measures the minute electric currents generated by sulphide deposits.

Semi-autogenous grinding (SAG) – A method of grinding rock into fine powder whereby the grinding media consist of larger chunks of rocks and steel balls.

Semivariogram — In geostatistics, a graph or mathematical function expressing the increase in variance of sample grades with increasing distance between sample locations.

Serpentine – A greenish, metamorphic mineral consisting of magnesium silicate.

Shaft – A vertical or inclined excavation in rock for the purpose of providing access to an orebody. Usually equipped with a hoist at the top, which lowers and raises a conveyance for handling workers and materials.

Shale – Sedimentary rock formed by the consolidation of mud or silt.

Shear or shearing – The deformation of rocks by lateral movement along innumerable parallel planes, generally resulting from pressure and producing such metamorphic structures as cleavage and schistosity.

Shear zone – A zone in which shearing has occurred on a large scale.

Sheave wheel – A large, grooved wheel in the top of a headframe over which the hoisting rope passes.

Shoot – A concentration of mineral values; that part of a vein or zone carrying values of ore grade.

Short selling – The borrowing of stock from a broker in order to sell it in the hope that it may be purchased at a lower price later on.

Short ton – 2,000 lbs. avoirdupois.

Showing – A mineral occurrence that has been located, but whose extent is not known.

Shrinkage stoping – A stoping method which uses part of the broken ore as a working platform and as support for the walls of the stope.

Siderite – Iron carbonate, which when pure, contains 48.2% iron; must be roasted to drive off carbon dioxide before it can be used in a blast furnace. Roasted product is called sinter.

Silica – Silicon dioxide. Quartz is a common example.

Siliceous – A rock containing an abundance of quartz.

Sill – An intrusive sheet of igneous rock of roughly uniform thickness that has been forced between the bedding planes of existing rock.

Silt – Muddy deposits of fine sediment usually found on the bottoms of lakes.

Sinter – Fine particles of iron ore that have been treated by heat to produce blast furnace feed.**Skarn** – Name for the metamorphic rocks surrounding an igneous intrusive where it comes in contact with a limestone or dolostone formation.

Skip – A self-dumping bucket used in a shaft for hoisting ore or rock.

Slag – The vitreous mass separated from the fused metals in the smelting process.

Slash – The process of blasting rock from the side of an underground opening to widen the opening.

Slate – A metamorphic rock; the metamorphic equivalent of shale.

Slickenside – The striated, polished surface of a fault caused by one wall rubbing against the other.

Sludge – Rock cuttings from a diamond drill hole, sometimes used for assaying.

Sodium cyanide – A chemical used in the milling of gold ores to dissolve gold and silver.

Solvent extraction-electrowinning (SX-EW) – A metallurgical technique, so far applied only to copper ores, in which metal is dissolved from the rock by organic solvents and recovered from solution by electrolysis.

Specific gravity – Density.

Spelter – The zinc of commerce, more or less impure, cast from molten metal into slabs or ingots.

Sphalerite – A zinc sulphide mineral; the most common ore mineral of zinc.

Split – The shareholder-approved division of a company's outstanding common shares into a larger number of new common shares. The opposite is a share consolidation.

Spot price – Current delivery price of a commodity traded in the spot market.

Station – An enlargement of a shaft made for the storage and handling of equipment and for driving drifts at that elevation.

Step-out drilling – Holes drilled to intersect a hoped-for continuation of a mineralized horizon or structure along strike or down dip.

Stock exchange – An organized market concerned with the buying and selling of common and preferred shares and warrants by stockbrokers who own seats on the exchange and meet membership requirements.

Stockpile – Broken ore heaped on surface, pending treatment or shipment.

Stockwork – A pipe- or funnel-shaped zone of fracturing, where veins, veinlets, and disseminated mineralization tend to form. Common in the footwall of massive sulphide deposits and in mineralization around intrusive rocks.

Stope – An excavation in a mine from which ore is, or has been, extracted.

Stop-loss order – An arrangement whereby a client gives his broker instructions to sell a stock if and when its price drops to a specified figure on the market.

Stratigraphy – Strictly, the description of bedded rock sequences; used loosely, the sequence of bedded rocks in a particular area.

Streak – A diagnostic characteristic of minerals, where scratching a sample on a piece of unglazed porcelain leaves powder of a characteristic color.

Street certificate – A certificate representing ownership in a specified number of shares that is registered in the name of some previous owner who has endorsed the certificate so that it may be transferred to a new owner without referral to transfer agent.

Striations – Prominent parallel scratches left on bedrock by advancing glaciers.

Strike – The direction, or bearing from true north, of a vein or rock formation measured on a horizontal surface.

Stringer – A narrow vein or irregular filament of a mineral or minerals traversing a rock mass.

Strip – To remove the overburden or waste rock overlying an orebody in preparation for mining by open pit methods.

Stripping ratio – The ratio of tonnes removed as waste to the number of tonnes of ore removed from an open-pit mine.

Strip mine – An open-pit mine, usually a coal mine, operated by removing overburden, excavating the coal seam, then returning the overburden.

Sub-bituminous – A black coal, intermediate between lignite and bituminous.

Sublevel – A level or working horizon in a mine between main working levels.

Subsidiary company – A company in which the majority of shares (a controlling position) is held by another company.

Sulphide – A compound of sulphur and some other element; most base-metal ore minerals are sulphides.

Sulphide dust explosions – An underground mining hazard involving the spontaneous combustion of airborne dust containing sulphide minerals.

Sulphur dioxide – A gas liberated during the smelting of most sulphide ores; either converted into sulphuric acid or released into the atmosphere in the form of a gas.

Sulphide ore – An ore composed mainly of sulphide minerals; in gold deposits in tropical or arid regions, the primary ore in unweathered rock below the zone of weathering (see "oxide ore").

Sump – An underground excavation where water accumulates before being pumped to surface.

Supergene – Processes occurring close to the Earth's surface.

Supergene enrichment – A process by which mineralization is enriched by the circulation of groundwater and the weathering process; significant in porphyry-copper and iron oxide-copper-gold deposits, where zones of much higher-grade mineralization may be formed.

Sustainable development – Industrial development that does not detract from the potential of the natural environment to provide benefits to future generations.

Syenite – An intrusive igneous rock composed chiefly of orthoclase.

Sylvite – potassium chloride, the principal ore of potassium mined for fertilizer manufacturing.

Syncline – A down-arching fold in bedded rocks.

Synform – A fold similar to a syncline, but where relative ages of the folded rocks are unknown.

Syngenetic – A term used to describe when mineralization in a deposit was formed relative to the host rocks in which it is found. In this case, the mineralization was formed at the same time as the host rocks. (The opposite is epigenetic.)

T

Taconite – A highly abrasive iron ore.

Tailings – Material rejected from a mill after most of the recoverable valuable minerals have been extracted.

Tailings pond – A low-lying depression used to confine tailings, the prime function of which is to allow enough time for heavy metals to settle out or for cyanide to be destroyed before water is discharged into the local watershed.

Talus – A heap of broken, coarse rock found at the base of a cliff or mountain.

Telluride – A chemical compound consisting of the element tellurium and another element, often gold or silver.

Thermal coal – Coal burned to generate the steam that drives turbines to generate electricity.

Thickener – A large, round tank used in milling operations to separate solids from liquids; clear fluid overflows from the tank and rock particles sink to the bottom.

Ton unit — In metallurgy, a unit of mass equal to one per cent of a ton; thus if an ore assays 3% metal, one ton is said to hold 3 ton units. The "unit" may be based on contained metal, or on the contained oxide or other compound. A short ton unit is equal to 20 lb., a long ton unit, 22.4 lb., and a metric ton unit (or "tonne unit"), 10 kg. Occasionally used in smelting contracts, or in pricing some ores.

Tonne – The metric ton, equal to 1,000 kilograms.

Tonnes-per-vertical-metre – Common unit used to describe the amount of ore in a deposit; ore length is multiplied by the width and divided by the appropriate rock factor to give the amount of ore for each vertical metre of depth.

Train — The "plume" of soil or debris created by mechanical movement of surface materials — for example, the scouring of bedrock by glaciers. Boulder or indicator-mineral trains may lead prospectors back to mineral occurrences.

Tram – To haul cars of ore or waste in a mine.

Treasury shares – The unissued shares in a company's treasury.

Trench – A long, narrow excavation dug through overburden, or blasted out of rock, to expose a vein or ore structure.

Trend – The direction, in the horizontal plane, of a linear geological feature, such as an ore zone, measured from true north.

Tube mill – An apparatus consisting of a revolving cylinder about half-filled with steel rods or balls and into which crushed ore is fed for fine grinding.

Tuff – Rock composed of fine volcanic ash.

Tunnel – A horizontal underground opening, open to the atmosphere at both ends.

Tunnel-boring-machine – A machine used to excavate a tunnel through soil or rock by mechanical means as opposed to drilling and blasting.

U

Umpire sample or assay – An assay made by a third party to provide a basis for settling disputes between buyers and sellers of ore.

Uncut value – The actual assay value of a core sample as opposed to a cut value which has been reduced by some arbitrary formula.

Undercut-and-fill — a cut-and-fill mining method that works downward, with cemented fill placed above the working area; best suited for poor ground conditions.

Underwrite – A firm commitment made by a broker or other financial institution to purchase a block of shares at a specified price.

Unlisted – A stock not traded on a recognized exchange or quotation system.

Uraninite – A uranium mineral with a high uranium oxide content. Frequently found in pegmatite dykes.

Uranium – A radioactive, silvery-white, metallic element.

V

Vein – A fissure, fault or crack in a rock filled by minerals that have travelled upwards from some deep source.

Vendor – A seller. In the case of mining companies, the consideration paid for properties purchased is often a block of treasury shares. These shares are termed vendor shares and are normally pooled or escrowed.

Visible gold – Native gold which is discernible, in a hand specimen, to the unaided eye.

Volcanic rocks – Igneous rocks formed from magma that has flowed out or has been violently ejected from a volcano.

Volcanogenic – A term used to describe the volcanic origin of mineralization.

Voting right – The stockholder's right to vote in the affairs of the company. Most common shares have one vote each. Preferred stock usually has the right to vote when preferred dividends are in default.

Vug – A small cavity in a rock, frequently lined with well-formed crystals. Amethyst commonly forms in these cavities.

W

Wall rocks – Rock units on either side of an orebody. The hangingwall and footwall rocks of an orebody.

Warrant – See Rights.

Waste – Unmineralized, or sometimes mineralized, rock that is not minable at a profit.

Water gel — an explosive mixture containing a solution of oxidant (usually ammonium nitrate) in water plus an explosive.

Weathering - The destruction or alteration of primary minerals in a rock; also, erosion.

Wedge – A technique of directing a diamond drill hole in a desired direction away from its current orientation.

Winze – An internal shaft.

Witness post – A claim post placed on a claim line when it cannot be placed in the corner of a claim because of water or difficult terrain.

Work index — The quantity of energy used in crushing or grinding a material to a specified grain size; usually expressed in kilowatt-hours per tonne. Soft material would typically show a grinding work index below 9 kWh/t, while that of very hard ores may exceed 20. Crushing work indices range from 5 to 40 kWh/t.

Working capital – The liquid resources a company has to meet day-to-day expenses of operation; defined as the excess of current assets over current liabilities.

Writeoffs – Amounts deducted from a company's reported profit for depreciation or preproduction costs. Writeoffs are not an out-of-pocket expense, but reduce the amount of taxable profit.

X

Xenolith – A fragment of country rock enclosed in an intrusive rock.

Y

Yield – The current annual dividend rate expressed as a percentage of the current market price of the stock.

Z

Zone – An area of distinct mineralization.

Zone of oxidation – The upper portion of an orebody that has been oxidized.

Index

Appendix I

CONVERSION FACTORS FOR MEASUREMENTS

Conversion from SI to Imperial | Conversion from Imperial to SI

SI UNIT	MULTIPLIED BY	GIVES	IMPERIAL UNIT	MULTIPLIED BY	GIVES
		LENGTH			
1 mm	0.039	inches	1 inch	25.4	mm
1 cm	0.394	inches	1 inch	2.54	cm
1 m	3.281	feet	1 foot	0.305	m
1 km	0.621	miles (statute)	1 mile (statute)	1.609	km
		AREA			
1 m²	10.764	square feet	1 square foot	0.093	m²
1 km²	0.386	square miles	1 square mile	2.590	km²
1 ha	2.471	acres	1 acre	0.405	ha
		VOLUME			
1 cm³	0.061	cubic inches	1 cubic inch	16.387	cm³
1 L	0.264	gallon (U.S.)	1 gal	3.785	L
1 L	0.006	barrel (bbl)	1 bbl	159	L
1 m³	1.308	cubic yards	1 cubic yard	0.765	m³
		MASS			
1 g	0.035	ounces (avdp)	1 ounce (avdp)	28.350	g
1 g	0.032	ounces (troy)	1 ounce (troy)	31.103	g
1 kg	2.205	pounds (avdp)	1 pound (avdp)	0.454	kg
1 t	32,151	ounces (troy)	1 ounce (troy)	0.000031	t
1 t	1.102	tons (short)	1 ton (short)	0.907	t
1 t	0.984	tons (long)	1 ton (long)	1.016	t
		CONCENTRATION			
1 g/t	0.029	ounce (troy)/ ton (short)	1 ounce (troy)/ ton (short)	34.286	g/t
		DENSITY, PRESSURE, ENERGY, POWER			
1 t/m³	62.428	lb/ft³	1 lb/ft³	0.016	t/m³
1 kPa	0.145	lb/in² (psi)	1 lb/in²	6.895	kPa
1 J	0.738	ft-lb	1 ft-lb	1.356	J
1 kJ	0.948	BTU	1 BTU	1.055	kJ
1 kW	1.341	hp	1 hp	0.746	kW

SOME USEFUL UNITS

1 ct = 0.2 g

1 ppm = 1 g/t = 0.0001%

1 ppb = 0.001 g/t

1 t of gold = 32,151 oz. gold

Appendix II

PERIODIC TABLE OF THE ELEMENTS

Periodic Table of the Elements

Group ⇨	1	2	3	4	5	6	7	8	9	10	11	12	13	14	15	16	17	18
Period ⇩																		
1	1 H																	2 He
2	3 Li	4 Be											5 B	6 C	7 N	8 O	9 F	10 Ne
3	11 Na	12 Mg											13 Al	14 Si	15 P	16 S	17 Cl	18 Ar
4	19 K	20 Ca	21 Sc	22 Ti	23 V	24 Cr	25 Mn	26 Fe	27 Co	28 Ni	29 Cu	30 Zn	31 Ga	32 Ge	33 As	34 Se	35 Br	36 Kr
5	37 Rb	38 Sr	39 Y	40 Zr	41 Nb	42 Mo	43 Tc	44 Ru	45 Rh	46 Pd	47 Ag	48 Cd	49 In	50 Sn	51 Sb	52 Te	53 I	54 Xe
6	55 Cs	56 Ba	57- 71	72 Hf	73 Ta	74 W	75 Re	76 Os	77 Ir	78 Pt	79 Au	80 Hg	81 Tl	82 Pb	83 Bi	84 Po	85 At	86 Rn
7	87 Fr	88 Ra	89- 103	104 Rf	105 Db	106 Sg	107 Bh	108 Hs	109 Mt	110 Ds	111 Rg							

⇩

Lanthanide Series	57 La	58 Ce	59 Pr	60 Nd	61 Pm	62 Sm	63 Eu	64 Gd	65 Tb	66 Dy	67 Ho	68 Er	69 Tm	70 Yb	71 Lu
Actinide Series	89 Ac	90 Th	91 Pa	92 U	93 Np	94 Pu	95 Am	96 Cm	97 Bk	98 Cf	99 Es	100 Fm	101 Md	102 No	103 Lr

1	Hydrogen	H	33	Arsenic	As	65	Terbium	Tb	97	Berkelium	Bk
2	Helium	He	34	Selenium	Se	66	Dysprosium	Dy	98	Californium	Cf
3	Lithium	Li	35	Bromine	Br	67	Holmium	Ho	99	Einsteinium	Es
4	Beryllium	Be	36	Krypton	Kr	68	Erbium	Er	100	Fermium	Fm
5	Boron	B	37	Rubidium	Rb	69	Thulium	Tm	101	Mendelevium	Md
6	Carbon	C	38	Strontium	Sr	70	Ytterbium	Yb	102	Nobelium	No
7	Nitrogen	N	39	Yttrium	Y	71	Lutetium	Lu	103	Lawrencium	Lr
8	Oxygen	O	40	Zirconium	Zr	72	Hafnium	Hf	104	Rutherfordium	Rf
9	Fluorine	F	41	Niobium	Nb	73	Tantalum	Ta	105	Dubnium	Db
10	Neon	Ne	42	Molybdenum	Mo	74	Tungsten	W	106	Seaborgium	Sg
11	Sodium	Na	43	Technetium	Tc	75	Rhenium	Re	107	Bohrium	Bh
12	Magnesium	Mg	44	Ruthenium	Ru	76	Osmium	Os	108	Hassium	Hs
13	Aluminum	Al	45	Rhodium	Rh	77	Iridium	Ir	109	Meitnerium	Mt
14	Silicon	Si	46	Palladium	Pd	78	Platinum	Pt	110	Darmstadtium	Ds
15	Phosphorus	P	47	Silver	Ag	79	Gold	Au	111	Roentgenium	Rg
16	Sulphur	S	48	Cadmium	Cd	80	Mercury	Hg	Elements 112, 114, 116:		
17	Chlorine	Cl	49	Indium	In	81	Thallium	Tl	Reported but not named		
18	Argon	Ar	50	Tin	Sn	82	Lead	Pb			
19	Potassium	K	51	Antimony	Sb	83	Bismuth	Bi			
20	Calcium	Ca	52	Tellurium	Te	84	Astatine	At			
21	Scandium	Sc	53	Iodine	I	85	Polonium	Po			
22	Titanium	Ti	54	Xenon	Xe	86	Radon	Rn			
23	Vanadium	V	55	Cesium	Cs	87	Francium	Fr			
24	Chromium	Cr	56	Barium	Ba	88	Radium	Ra			
25	Manganese	Mn	57	Lanthanum	La	89	Actinium	Ac			
26	Iron	Fe	58	Cerium	Ce	90	Thorium	Th			
27	Cobalt	Co	59	Praseodymium	Pr	91	Protactinium	Pa			
28	Nickel	Ni	60	Neodymium	Nd	92	Uranium	U			
29	Copper	Cu	61	Promethium	Pm	93	Neptunium	Np			
30	Zinc	Zn	62	Samarium	Sm	94	Plutonium	Pu			
31	Gallium	Ga	63	Europium	Eu	95	Americium	Am			
32	Germanium	Ge	64	Gadolinium	Gd	96	Curium	Cm			

Appendix III

ROCK CLASSIFICATION

Table of Igneous Rocks

Composition	Mineralogy	Intrusive (Plutonic) Rocks	Etrusive (Volcanic) Rock
Felsic (55-75%) silica	Quartz and feldspar	Granite	Rhyolite
	Mainly feldspar	Syenite	Trachyte
Intermediate-Felsic (55-65% silica)	Mainly potassium feldspar	Monzonite	Latite
	Mainly sodium feldspar	Granodiorite	Dacite
Intermediate (50-60% silica)	Sodium feldspar	Diorite	Andesite
Mafic (45-50% silica)	Feldspar and ferromagnesian minerals	Gabbro, diabase	Basalt
Ultramafic (under 45% silica)	Ferromagnesian minerals	Peridotite	Komatiite

Table of Sedimentary Rocks

Class	Composition	Rock Type
Chemical Sediments	Mainly calcite	Limestone
	Mainly dolomite	Dolostone
	Fine-grained quartz	Chert
	Iron minerals	Iron formation
	Mineral salts	Evaporite
Clastic Sediments	Gravel and cobbles (over 4mm)	Conglomerate
	Sand (0.07-4 mm)	Sandstone
	Sand (plus fines)	Graywacke
	Fines (silt and clay)	Shale (siltstone, mudstone)

Table of Metamorphic Rocks

Grain Size	Rock Type	Characteristics
Coarse	Gneiss	Strongly foliated, hard rock
	Amphibolite	Like gneiss, but mainly dark minerals
	Schist	Strongly foliated, but easily split
Fine	Phyllite	Strongly foliated, but fine-grained
	Slate	Foliation less developed, fine-grained
Variable	Hornfels	Unfoliated
Coarse	Marble	Calcite and dolomite
	Quartzite	Coarse quartz

Appendix IV

Geological Time Scale

Eon	Era	Period	Age (Million yrs)
Phanerozoic	Cenozoic	Quaternary	Present 1.65
		Tertiary	65
	Mesozoic	Cretaceous	145
		Jurassic	200
		Triassic	251
	Paleozoic	Permian	300
		Pennsylvanian / Carboniferous (Europe) / Mississippian	314
			355
		Devonian	418
		Silurian	441
		Ordovician	490
		Cambrian	544
Proterozoic or Late Precambrian	Late Proterozoic		1000
	Middle Proterozoic		1750
	Early Proterozoic		2500
Archean or Early Precambrian			4600

Appendix V

Metals and their Ore Minerals

Metal	Common Ore Minerals
Aluminum	Bauxite (hydrated aluminum oxides)
Cobalt	Colbaltite (cobalt sulpharsenide, 36% Co)
Chromium	Chromite (ferrous chromic oxide, 46% Cr)
Copper	Native Copper
	Chalcopyrite (copper iron sulphide, 35% Cu)
	Chalcocite (copper sulphide, 80% Cu)
	Bornite (copper iron sulphide, 63% Cu)
Gold	Native Gold
Iron	Hematite (iron oxide, 70% Fe)
	Magnetite (iron oxide, 72% Fe)
	Siderite (iron carbonate, 48% Fe)
Lead	Galena (lead sulphide, 87% Pb)
Molybdenum	Molybdenite (molybdenum disulphide, 60% Mo)
Nickel	Pentlandite (nickel iron sulphide, 22% Ni)
Silver	Native Silver
Tin	Cassiterite (tin oxide, 79% Sn)
Titanium	Ilmenite (iron titanium oxide, 32% Ti)
Tungsten	Wolframite (iron magnesium tungstate, 77% WO_3)
	Scheelite (calcium tungstate, 81% WO_3)
Uranium	Pitchblende (uranium oxide, 50-58% U_3O_8)
Zinc	Sphalerite (zinc sulphide, 67% Zn)

Appendix VI

SHIELDS AND FOLDBELTS

Foldbelts 0 - 200 Million Years Old

Foldbelts 200 - 570 Million Years Old

Precambrian Rocks Older Than 570 Million Years

After Duncan R. Derry

Appendix VII

TECTONIC PLATES

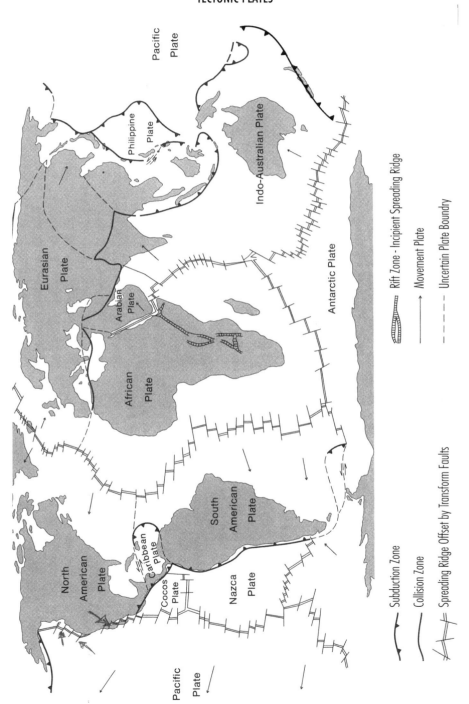

After Duncan R. Derry

Appendix VIII

INDUSTRY CLUSTERS

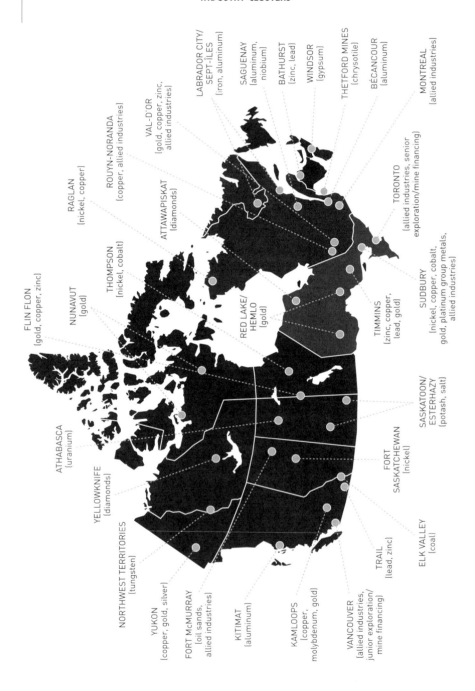

Common Rock Types

Photographs provided by the University of Toronto

Granite

Porphyry

Kimberlite

Basalt

Tuff

Conglomerate

Sandstone

Siltstone (showing laminations)

Limestone

Gneiss (with deformed foliation)

Typical Ore Minerals

Photographs provided by the University of Toronto

Quartz vein in
volcanic rock

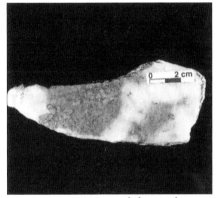

Native gold and
pyrite in quartz vein

Pyrite in quartz vein

Chalcopyrite

Pyrrhotite, chalcopyrite, pentlandite

Galena (grey) and sphalerite (brown)

Sphalerite and chalcopyrite

Hematite

Magnetite

Asbestos seam in serpentine peridotite